it

The Architecture of Existence

it

The Architecture of Existence

Gary

Volume Three

JANUS PUBLISHING COMPANY
London, England

First published in Great Britain 2008
by Janus Publishing Company Ltd,
105-107 Gloucester Place,
London W1U 6BY

www.januspublishing.co.uk

British Library Cataloguing-in-Publication Data
A catalogue record for this book is available from the British Library

Paperback ISBN Vol. 1. 978-1-85756-648-2
Paperback ISBN Vol. 2. 978-1-85756-653-6
Paperback ISBN Vol. 3. 978-1-85756-658-1
Paperback ISBN Vol. 4. 978-1-85756-633-8
Hardback ISBN 978-1-85756-668-0

Cover Design: Gary Hansen

Printed and bound in Great Britain

To whom **it** may concern

Contents

Contents

Introduction to Volume Three

The third of four volumes, *Volume Three, Part IV – Volition,* raises the stakes for those who achieve a state of consciousness whereby opportunities for self-improvement arise through decision-making, but not without risk of catastrophic failure!

The opening of the mind to increased possibilities carries heavy responsibilities to forestall misadventure. Conflict between competitive interests demands that man's ability to reason, having been acquired, be applied with unwavering resolve that the benefit of doubt always be weighted to deny detrimental effects.

Mankind is a caretaker but he does not always subscribe to his as-yet-unwritten code of caring behaviour. He still swings in his decision-trees from one extremity to the opposite. The higher the potentials are for raising himself up, the greater are the prospects of a calamitous fall. Therein lies his Achilles' heel: his vulnerability to self-induced misfortune. Volition is volatile!

The key to acting in his best interests lies in his ability to discriminate in favour of his best interests. The key to safeguarding his best interests resides in his ability to make rules and to enforce them rigorously while remaining open to opportunities for improvement; all of which point towards his affections being his ultimate arbiters.

The final volume, *Volume Four,* addresses the affections.

Part IV

VOLITION

106 Will: The Lever of Probability

Man has acquired consciousness and intellect and through that process he has assumed a capacity to direct his energies towards attaining goals. We now examine the nature of that director faculty: the will.

Will as Probability

Before consciousness, events unfolded strictly in accordance with physical principles. Forces resolved in the manner we have already shown, the stronger forces commanding the weaker forces proportionately. 'What', 'where' and 'when' were thus determined by the unwritten rules of natural tendency. 'Will-as-probability' governed.

As an understanding of these processes evolved in man he was able to pose the question, 'why?' Implicit in the question 'why?' is the inference that there may be alternative answers. Which answer applied to any specific question was open to question. If open and pending closure, man could raise a second question: 'what if?' What if he intervened, lending his personal strength to bias the resolution of forces in his own favour? What he was doing in entertaining such an idea was subtly converting the word 'will' in 'probability', to 'will' in 'greater probability'. By the grace of man, man found that he was able to influence the 'whats', 'wheres' and 'whens' of his immediate circumstances. Thus, what otherwise would have happened had he not intervened, will no longer happen. Instead, what happens comes to be the will of his determination.

Improbability

We use the term 'improbable' either in the context of odds against an event occurring or where we register surprise at a revelation. The forces that govern an improbable event are, in order of authority, physical, chemical and biological. We expect gravity to command mass; we are familiar with a multitude of chemical reactions between elements and compounds and we are becoming increasingly aware of complex behavioural characteristics of biological organisms, yet we are frequently surprised by the nature of new knowledge. Upon first acquaintance the common bath sponge exhibits much the same cellular qualities as a piece of pumice or lava, but with the added attribute of subtlety. On learning that sponge is or was alive we generally assume from its surface-attached seabed habitat and growth characteristics that it is an aquatic plant. When we acquire the improbable truth that it is a multi-cellular, water-pumping, filtering, micro-organism-gobbling animal, the improbability factor is notched up considerably. But the sponge hasn't finished with you yet! When a sponge is ground into small fragments and introduced into a conducive solution, it will reassemble itself into a sponge. How improbable is that? Who is driving what, where and how? Apparently 'where there is a "will" there is a way!'

Will as Intent

The drifts in meaning from will 'probably' to 'will more probably' to 'will' meaning 'shall with certainty' are excusable because 'meanings' are subjective. Each subject determines in his own mind what will probably happen if he does not intervene, what will likely happen if he does, and what he has to do to make a favourable outcome happen.

The fact that the meaning of 'will-as-intent' is subtly different from the meaning of 'will-as-probability' is significant. The former implies resolve; however, the threshold of that distinction is fuzzy. If we define 'will-as-intent' to be the desire for an outcome in one's perceived best interests, then such a desire can be construed to be a natural tendency. If we define will as natural tendency, as when sediments (will) tend to build up at a bend in a river, we cannot construe such will as intent unless we attach an animate quality to the sand or the water, or attribute such occurrence to the will of a greater authority, nature, the universe or God, for which we have no more rational evidence of 'will-as-intent' than we have for sand. So we frequently use these two concepts of 'will-as-intent' and 'will-as-probability' interchangeably, so habitually that we are persuaded that the inanimate is somehow animate and that the universe 'intends'. And who are we to deny that the universe or God is not animate and therefore wilful, since we know in the pragmatic, objective sense so little about them. Therefore we choose (will) to believe or not in such animation or higher order of 'will' to the extent that it serves our perceived best interests or serves to explain events. Today's consommé 'will' become tomorrow's consummate 'will'!

Willy-Nilly

Having said so, we have established a condition whereby the will has become captive, not of a law that prescribes conditions under which we use the word to connote 'intention' or 'natural tendency', but of a law-less condition enabling us to interchange meanings 'at will', without accountability. A simple semantic economy of means has led us to an excruciating conceptual polarity of ends. The consequence of such a rift in interpretation of the term 'will' naturally and inevitably predisposes us towards science and religion. We may call white black and black white, indeed give life to the lifeless of conventional wisdom, and vice versa, at will!

Understanding this predicament, and wishing to bring order to bear where apparent intellectual chaos reigns, it would seem appropriate to agree upon one of three courses. The first would be to make clear linguistic rules as to when one uses the word 'will' to signify intent and when to signify natural tendency, the distinction being established by the presence or absence of consciousness. The second: to accord to all things an inanimate state; and the third, to ascribe to all things a soul or life force, each with its own distinct natural tendencies or predispositions that arise from its pedigree origins and experience, treating each as if it has 'desire' to be, and to continue to be, itself, without interference, and to exercise its own 'will' according to 'its' own natural disposition. Have we not come to a point where 'will' cannot be defined as the antithesis of automatism since will and automatism both respond to what the secular would call fate and the cleric would describe as the will of God, neither of which we can test or prove?

The Will to Want to Be

If we credit all phenomena with animate identities, then we can say that so-and-so is performing according to its intrinsic nature, that is, as it must or 'wants to be'. This 'wanting' is no abstraction, it is an urgent matter of compelling necessity; it is will. It is the 'will' of rubber, for example, to return to its original form after being subjected to compression or tension; to such 'will' we accord the quality 'memory'.

It is the 'will' of sunflowers to face the sun. It is the 'will' of heat to dissipate and so on. It is the 'will' of all things to continue what they are doing, naturally and without interference, a quality that we call inertia. Whether the will belongs to the object, to nature, to the universe or to God is of scant significance. What is important is that there is this driving inclination to be or do something, regardless of consciousness.

By its nature, or more correctly its relation to nature, a pile of cinders 'wants to be' a cone, and that is the form that a volcano takes. Its constituent particles demonstrate 'an agreement in principle' to compromise their individual 'wills' to reach the centre of gravity of the earth by the most direct route. One might almost say that, through sharing a common 'goal', they have been able to forge a cohesive unit of collective purpose, a 'social compact', dense both in mass and in constitutional simplicity. Density of mass ensures that its 'will' will generally prevail, constitutional density assures that it is not easily distracted from its intended purpose. With this authority, the mountains are here to stay for a long while. Volcanoes are republican in the archaic sense; that is, they are commonwealth states of beings acting collectively in their combined best interests to be and endure.

By contrast a plant has a more complex set of 'goals'. It reaches downward to seek necessary sustenance from the soil in order to maintain its state of equilibrium and to sustain its simultaneous advance upward in search of the energy needed to drive its turbines of vitality. A plant's 'will' is more subtle than that of a stone. It has an ulterior motive, an unwritten agenda to advance its destiny of purpose. What destiny is this, more 'best interests to be and endure'? Indeed, if not in self, then, with generosity of spirit, in kind. What is this 'generosity' if a plant has no consciousness? It is its second nature to give without counting the cost, to serve without heeding the wounds, and, when all is played out, to remit itself ultimately and totally to the nourishment and support of others: how generous, how reciprocal, how democratic, how alive, how natural!

The animal combines the two, the determined mind of the mineral and the self-reproducing, nurturing soul of the plant, greatly facilitated by the gift of mobility: what a contradiction, what an aberration, what a disaster, what a delight!

These material and immaterial essences, mineral, plant and animal, to which is added the hybrid attributes of conscience and volition, coalescing out of the mix of parts, comprise the younger or should we say less mature but infinitely more complex earthly resident, man. The human, being introspective, 'wants to be' forever. This goal, the ego-ideal, conditions his will and behaviour and leads him towards a fate that is far greater in potential but far less certain than that of his forebears.

The Origins of Will

Let us take a moment to consider what conditions enabled conscious will to come into being. If we look backward in evolution we must necessarily concede that our forebears had conscious will, as did theirs. The question simply stated reduces to how far we are willing to look back for the origins of will, to the Stone Age or to stones? We are quick to concede that conscious will evolved gradually. That process appears to have been so gradual that we cannot recognize a quantum step between will-as-probability and will-as-intent. We may be willing to concede that a stone has no qualities that suggest to us that it has any vestiges of self-determination, but how about Stone Age man? Chimpanzees? Mammals? Vertebrates? Invertebrates? Amoebas?

A hasty return to the present will confirm that mankind is hardly insulated from the effects of will-as-probability. What we are experiencing is the selective leverage of probability whereby we intentionally enhance our prospects.

The Will to Foresee

Consider the case of the human being. Man has a myriad of influences bearing upon him, not the least of which are his prior experiences represented by imperfect memories unique to himself, in addition to his material, spatial and temporal states at any moment in time. If all the criteria by which man is affected at a particular moment could be known and quantified, a condition that would appear to be impossible since they are constantly changing, then, we may have a model that operates strictly according to mechanistic principles espoused by Descartes and Laplace. Their now somewhat disdained theories would be vindicated. There is no way of forecasting whether this could ever be achieved, and therefore no way to confirm or deny the efficacy of their theories with impeccable authority. We must settle for the improbable course of enhancing probability within the limits imposed by possibility.

Improbable Will

Probability only suggests what will more likely unfold from a set of alternatives. Probability does not entertain the prospect of future alternatives that do not exist 'naturally'. The advent of 'will-as-intent' heralds a new kind of uncertainty, not novelty that may arise due to accidental relations, as products of the puff, slosh and grind of our worldly mixing machine, but artificial, 'manufactured' novelty. Thus, the advent of human 'will' ushers in a new component of uncertainty with the ability to override to some degree the predictability offered by otherwise natural laws of probability. On the one hand, human 'will' biases events in favour of benefits applicable to those who exercise such 'will' power, while, on the other hand, the effects upon others may or may not be beneficial to them. In consequence it becomes necessary either for each to exert his own 'will' in the pursuit of his own interests, or for some agreement to be struck between interested parties in order not to damage the interests of any party. Thus, we witness an increase in conflicts of interest vis-à-vis the wills of individuals, and with respect to group and collective interests.

Will in the Singular

The greater the number of parties at interest, the greater the prospects for conflict. The greater the conflict, the greater the degree of uncertainty and departure from a course of events that would otherwise likely occur 'naturally' due to probability. By this measure humanity moves towards a form of anarchy whereby it reverts to or exacerbates the laws of the jungle, laws that prescribe that, all other things being equal, superior force prevails where man is in conflict with man. What happens when 6 billion 'wills' exert force against each other is a cascading improbability that the outcome will be beneficial for the average individual.

The Collective Will

It therefore behoves us to agree upon ways and means by which to settle disputes short of resort to the use of force. That eventuality, the will to be civil, heralds the birth of civilization. Social change only occurs when society is ready for it. Society is only ready for such change when it is hurting enough.

Thus, man demonstrates through his collective will his desire to be restored more closely to the natural law as the most probable cause of and course for his own satisfaction. His most fundamental interests are coincidental to his natural affinities.

Liberation in Deliberation

Volition is not simply freedom of expression; it is liberation from the dictates of the automatism of the animal world. No longer are we ritualistic animals, but a new class of life. Deliberate life has evolved.

However, we have in the word 'will' two seemingly opposed ideas: first the notion that events 'will' inevitably occur and, second, that they are somehow subject to volition through the exercise of 'will'. This contradiction reopens the question as to whether 'will' is the antithesis of automatism or whether automatism is, of itself, an expression of 'will'. If we draw the conclusion that automatism 'is will', then we must enquire whether the converse is true: is 'will' automatic and therefore not freedom of choice as we like to believe it to be? If so, we are forced to reacquaint ourselves with the idea that future events are simply current or past events waiting to happen, and that they are as certain as present and past events without the necessity for belief in fate.

The Genie in a Bottle

We are the only animals on earth that can overrule our genes through the exercise of will that arises from intellect, distinguished from will as instinct. Nature doesn't care what humans do through the exercise of will or the expression of ignorance. Time is hers. All is recyclable on her terms. It is only those that live that suffer. With this realization we understand that it is only life that we serve, and insofar as life is dependent upon the non-living for support, our suffering increases proportionately to the severity of our loss of these resources.

Human will is to change as imagination is to thought; a variable-flow control mechanism that regulates the expenditure of resources in the interest of securing advantage. Will conditions states of belief and resolves to initiate action. Like temperature, will enjoys a high and influential status in the hierarchy of causes that enables certain events to occur and others to be denied; that is to say, will is a priority sequencer. The exercise of human will gives rise to prospects for acceleration in the rates of recycling and matching elements, compounding their affinities to generate novel organizations. The causes are perceived needs; the effects potential benefits.

107 Resolution: Deciding to Decide

The connection between 'will' and 'decision' is resolve. A will that harbours preference without intention to act is merely desire.

What 'act' is this of which we speak if not the act or process of reducing to a simpler form? Resolution suggests itself as a means of clarification of intent whereby all considerations except the course of action intended are set aside. Resolution may thus be viewed as the act of committing the will to execute a decision.

In order to elucidate the distinction between those that successfully act in a chosen direction and those whose actions are relatively ineffectual, even if they may be more gifted or virtuous, we need to examine the elements of resolution. The distinction between the two may be represented by a comparison between focus and eccentricity. In focus all forces are directed towards a single point, in this case a goal. In eccentricity forces are dissipated over a larger area thereby exerting less influence upon the point in question and reducing the prospects of overpowering a competitive force.

Based upon one's relative values, if one should decide to decide upon a particular goal, it follows that all the constituent parts necessary for its execution – intention, determination, tenacity, fortitude and perseverance – require to be focused upon that end.

Ultimately, what is important is deciding what is important and/or neglecting what is not.

108 Decision: The Choice Act

No Choice

Lack of opportunity translates to submission to chance, the embrace of uncertainty and an invitation to stress.

Single Choice

Through the exercise of will, the act of choice dilutes uncertainty and the stresses that attend. When we examine the nature of choice we discover that all decisions are identical; each switches alternative futures on or off. Acceptance or rejection are the only options!

There will be but one future within which all events will fit. Each answer reduces to a determination as to which prospective path is intended to come to pass.

Multiple Choice

Compound decisions may not be a succession of simple choices. Underlying many decisions is a need to know. There are those things that are known (relative truth), those that are known to be unknown (relative ignorance) and unknown unknowns (absolute ignorance). In addition, underlying intentions may be honourable, purposefully directed, disinterested, unconsciously prejudiced or deceitful.

Thus, the greatest potential danger in making multiple-choice decisions arises from not seeing beyond the range of obvious options and thereby presuming that all questions, and by implication the range of all possible answers, are clear and apparent. Further, we may elect to make a single positive decision (like to paint the front door vermilion red) or a series of negative decisions (like eliminating all colour options one by one until vermilion red remains the only option left).

Finally, the probability of making cumulative judgements that, when combined, deliver the optimal outcome are in inverse proportion to the number of alternatives available to choice, all other considerations being equal. So there you have it, and there you (probably) don't!

Temperate Choices

Regarding temperature and temperament: 'hot heads' act before they think, 'cold heads' over-deliberate, allowing events to overtake them. Somewhere between, in the temperate zones, resides an approximate balance between thought and action. This is the climatic zone where the great preponderance of development in the arts and sciences has occurred. It could, of course, be argued that Mesopotamia and the Nile Valley are not exactly temperate regions. But recalling that the dawn of civilization occurred at a time closer to the recession of the ice age, these regions, with their hanging gardens and groves of acanthus trees upon which the stone columns of Karnak were modelled, were more temperate than they are today.

Yeah or Nay

We may say 'yes' to something and 'no' to another. 'Yes' or 'no' is all there is to making decisions. There are no maybes. Of course, there are costs and benefits that we may elect to weigh against each other, but ultimately we must open or close the door to the flow of particular futures. There are no partial futures.

When we reduce decision-making to such simple terms we can manage it more simply. So simply, in fact, that we have relegated many of our decisions to automation and defer to machines to sort out our 'best interests' and best decision-options for us. This constitutes the most powerful idea behind the electronic computer; an ever-expanding battery of on-or-off switches that simply await our instructions: yeah or nay.

Go Climb a Tree

We can view decision-making as climbing a tree. At the outset all options are open. All parts are prospective destinations. But as we ascend we come to divides in the trunk or the limbs. We must state a preference for one and follow our line of choice. Some options may appear to remain open in that we may retrace our steps for a price and take alternative routes to alternative destinations, but others are denied as though the alternatives have been cut off with a chainsaw.

After a while we recognize that too many limbs have been lopped off. We cannot return to ground zero. The futures of the past are denied. We arrive in this way at the future of choice.

Conflicts of Interest

Our decisions are often impacted and modified by the decisions of others. This process gives cause for new decisions. Decisions that are not the products of 'free will'. We find our paths compromised by the fact that 'others' have climbed 'our' tree before us and decided to cut off some future options. We may even encounter such an 'other', chainsaw in hand, threatening to cut 'our' limbs if we do not accede to 'his' will and his decision. We may find that not only has our decision-tree been defaced, it has been denatured. No longer is there a prospect of gathering succulent sweet mulberries, but bitter, perhaps even poisonous fruit looms ahead: the fruit of more uncertainty and greater stress.

Compromise

Do you remember 'resolution', the step before making a decision? At this juncture neither party can make a decision without taking into account the interests of others. The prospect of another kind of decision looms, one made by agreement to reconcile difference.

Choice is seldom entirely 'free' but rather circumscribed by inclinations, inducements, constraints or compulsion. We discover that decision-making and stress are often mutually in attendance!

Lingering Doubts

Almost every decision involves speculation until the verity of its merits are borne out or denied by events over time. The truth or correctness of a decision presents itself as relative rather than absolute, lasting only until succeeding events present more advantageous alternatives, at a later date and in another context. We are not privy to knowledge of the future.

Conflicts of interest arise between the power to further advance personal interests and the power to convey benefits to others. In the ideal world we would make no decision that we knew in advance would be detrimental to the interests of anyone. In this state we would be neutral to, or abstain from, controversy. Ironically, that is the state that we ultimately achieve, but only when it is too late to make a conscious decision.

A Word About Voting

The most closely held secret about voting in a democracy is that everyone has more than one decision to make. A votive act consists of the expression of a desire. In electing not to vote a prospective voter is expressing such a desire, casting votes against candidates and/or issues *ipso facto* by virtue of being eligible to vote. In the context of multiple choice, a second type of decision involves determining which candidate(s) or issues to deny a vote to. Only upon elimination of all alternative options can a voter then determine the most favourable candidate or issue to vote for. Or is it so simple?

Let us take the hypothetical case where three or more candidates contend for a single office. Candidates *A* and *B* are in close contention for receiving the largest number of votes while candidate *C* appears to have a negligible chance of winning. For the purposes of this model we shall assume that all candidates have charisma and are well qualified to perform in the office to which they aspire.

As viewed by a single voter, you, the views of Candidate *A* are anathema. Under no circumstances would you vote for him. By comparison the views of Candidate *B* are moderate and acceptable. However, the positions of Candidate *C* on the issues more closely correspond to your own and are, in your judgement, enlightened as they portend for the future. The question of how you are going to address this conflict of interests requires two decisions: who are you going to deny your vote to, and only then who are you going to vote for. Are you going to cast a hate vote by voting for *B* who has the best prospect of stopping *A*, or a love vote for Candidate *C*?

In a democracy, almost without exception, the majority of eligible voters lose. In the typical election about 50 per cent of the electorate decline to vote. This, as we have shown, is tantamount to a no vote for all candidates. The winner receives the greatest number of yes votes cast for any single candidate. This, in a close two-way race, may be (say) 26 per cent of the total possible while his opponent receives 24 per cent. So losers outnumber winners by a ratio 74:26 or nearly 3:1. Thus, everybody elects but only a minority elects a winner!

Does it matter? The answer to this question has to do with who is affected by the results. If we are primarily interested in accruing gains and forestalling losses in the short term, then we tend to feel that our vote must count and therefore be counted amongst the winning votes. If, on the other hand, we are interested in shaping longer-term futures by weighting the balance of public debate in favour of moving towards ideals, then we may cast our vote on principle, even for a candidate with little chance of winning. How we vote will be determined subconsciously by our sense of relative security in our immediate and long-term futures, and by our sense of responsibility for others.

Decision Theory

In a world that has become progressively more complex arising from growing human populations, their processes, products and expanding inventories of information, the potential for decisions inadvertently generating detrimental effects has increased exponentially. In the interests of averting the worst of the worst possible outcomes, systematic approaches to decision-making have been developed. Decision Theory, a practice of modelling decision-contexts mathematically before the event of decision-making, breaks down and examines the elements of complexity as discrete topics, sequencing them to show the alternative outcomes of choice. By reconstituting the elements into modular building blocks optimal futures may be synthesized.

By such processes disparate issues including probabilities, values, sensitivity, uncertainty, risk, costs, benefits and non-monetary criteria may be ordered within logical structures.

The Delphi Method

Disparity of opinion compounds exponentially with an increase in the number of parties participating in decision-making. While the democratic process may work acceptably well for determining the directions that social issues may be allowed to evolve, democracy cannot be relied upon to optimize beneficial outcomes where life-critical issues are at stake.

In the interest of applying more scientifically defensible procedures to critical decision-making (distinquished from traditional theoretical, quantitative or linear-extension methods of predicting futures), the Delphi Method (a process oddly named after an ancient Greek oracle that infers the antithesis of the objective method sought) was developed by the RAND Corporation in the 1950s. The Delphi process calls upon experts to respond to questions regarding the likelihood and timing of hypothetical future events. Their conjectures are represented anonymously in the form of further questions in succeeding rounds, by which means the participants gravitate towards what are considered to be the most probable (distinquished from the most desirable) future events and their time frames. The final reductions are then not infrequently adopted as military/industrial goals and timetables.

Suffice to say and in any event – decisions tend to be decisive!

109 Habit: The Mantle of Disciples and Sloths

Habit

Habit, the personal inclination towards or acceptance of constancy, like inertia, extends the present into the future until affected by a force for change. Habits are neither good nor bad except to the extent that we interpret them to be as we read each situation in relation to our interests. One must assess costs against benefits to assess the implications of assuming habits. With respect to continuity in security of the mind, habit is a comfort; with respect to advancing greater security it is an impediment. Insofar as we need a measure of both, we 'inhabit' and 'exhabit' alternately.

Mantle

We are enveloped by habit. The extent to which habit works to our advantage is a function of degree. The mantle of habit governs appearances and restricts movement. How much is too much is often too much for many to assess. The real measure of adequacy is sufficiency in mental dexterity.

We are all prisoners of our mental states. We have survived our pasts and therefore see ourselves as would-be survivors under similar circumstances. Doubt harbours in our minds with respect to our abilities to survive futures under different circumstances.

Disciples

We have been disciplined by our pasts to believe that we should continue to survive in the manner that we have in the past. We become students of our successes and failures. We endeavour to optimize success and minimize failure. The conservative in each of us holds to steady states of being. However, there is no way today to deliver tomorrow by yearning for yesterday. Those of us that aspire to greater, more rewarding states of being, do so at our own risks. We are all disciples of our dispositions.

Having acceded to our predispositions, we ask ourselves how may we exploit our conditions to the greatest advantage? The answer comes to us, as it does to any disciple, with the application of discipline. Discipline may be self-imposed or imposed upon us by choice or circumstance. Those who are naturally disposed to exercise or submit to discipline are habitual disciples.

Sloths

Those who are not so disposed default through lack of energy or mental acuity. They are too lazy or slow in movement to effectively advance their interests and, as a result, allow events directed by external forces to govern their own conditions. They wear the cloth of the sloth. They are creatures of habit.

110 Principle: Accordance with Rules

Principle

Principle deals with patterns of behaviour with which we identify. In the absence of patterns we have what we call chaos, the appearance of randomness that forecloses our abilities to predict futures. Principle, like habit, is a system for mapping, in this case of values, whereby if we have the luxury of choice and a prospective destination is not on our map, we don't go there. If we did, we would feel insecure about our ability to function well or to return to our prior positions. One doesn't climb unclimbed mountains on principle; one finds other reasons.

The Rule of Rules

To act on principle is to adhere to a set of rules that has been formulated on the basis of experience or the intellect and has found acceptance. The question then arises, why should one conduct oneself similarly under similar circumstances? The answer lies in probability, the uncompromising strength of scientific method whereby, all things being equal, like causes give rise to like effects. So acting upon principle may be viewed as analogous to acting conservatively by reason of dependence upon precedence; and in the understanding that history not only repeats itself, but inevitably so in the context of natural law. There is, however, another aspect to principle that prompts us to place reliance upon untried, often self-formulated rules. Logic underlies accountability. We feel the need to account to ourselves and to others for our actions. In accounting we move to offset uncertainty with certainty. In resorting to logic we can defend ourselves credibly against challenges from reason or the emotions. Reason reconciles the incongruous.

Fundaments

The structures of such rules are so fundamental and so well established in logic that if they were held by geometers they would be called axioms. Some people make careers out of acting upon principle. An accountant accepts the system of enumeration of his culture as sacrosanct, while others conduct their businesses based upon other parameters, like personal satisfaction, whereby goods are bartered to the satisfaction of both parties without reliance upon logical derivatives. Both systems can be construed to be expressions of principle; their currencies are merely different. Both carry appearances of action in support of self-interest.

Assumption

Principle may be tailored like a coat to fit one's aspirations for personal comfort or social acceptance. Principle is thus the instrument that helps one to get from where one is to where one would like to be.

Acting upon principle carries with it a connotation of plausible confidence of bearing. One who acts on principle is seen by others to have direction. If, on the other hand, others wish to obstruct progress in that certain direction, they do not need to know very much about the subject, you; they simply need to recognize the principles that are being adhered to, and to undermine those principles. So acting on principle, on the one hand, facilitates progress, while, on the other hand, it makes the bearer vulnerable to the conflicting interests of others who can read his map and undermine his best intentions and efforts.

Options

Typically people don't operate on the basis of principle; pragmatism seems to serve them better in attaining short-term goals. However, principles constitute the operating systems of the universe that are responsible for the beautifully balanced relations inherent in nature. In departing from principle, man creates his own difficulties. Nature's operating systems effectively ensure long-term survival and integrity of the system as a whole while permitting attrition in the integrity of its parts, all of which are recycled.

111 Benefit: Good Fit

In recalling our earlier recognition that intelligence translated to insight into essential relationships, what we recognized there were inherent qualities of correctness of relation. In that context correctness was identifiable as 'good fit' in the sense of correctness of perception or assessment vis-à-vis the laws of physics.

Connotations of Value

When we examine the idea of 'benefit' we place a stringent condition upon our definition of goodness of 'fit': the necessity for relations to be in the perceived interests of those upon whom the relations bear. This is of course a narrower construct since it raises the question as to who is, or is to be, served by the prescription. Contention and competition arise in instances where benefits are not generally applicable or equitably shared. We must ask ourselves who the beneficiary or beneficiaries are to be. In cases where there is only one beneficiary, oneself, and particularly where no other parties are disadvantaged, the question appears to be moot and the answer self-evident. The value passes to he who has the singular interest.

Inferences of Conflict of Interest

Where two parties are in contention for a single benefit we discover that there is literal disagreement as to what 'good fit' means. A benefit that favours one at the expense of another is not so defined by the second party. It appears that the dictionary meaning of benefit is not consistent as between subjects but only with respect to objects. Thus, a loaf of bread is beneficial in principle, but in practice is only so for those that eat it.

The Physical Ideal

The prospect of sufficiency of supply to meet demands with respect to needs is idyllic, but it is effectual for the few but not for the many. Society is organized to improve supplies for the needy, but only conditionally, the condition being their ability to pay. This, it may be argued, is just in the sense that rewards are by definition compensation for service. But the logic of the argument becomes clouded where it pertains to necessities essential to the support of life and human dignity, particularly where the nature of the service to be rendered is unclear.

The Moral Ideal

Ah, dignity; we have introduced a new parameter to the equation of supply and demand, a parameter that relies upon mental rather than physical prowess. Not simply mental dexterity, but a special way of using intelligence that projects personal interest beyond the moment and beyond the self to other times and other people in need who, but for the grace of God, could be us.

Insofar as we are not privy to God's will and future intentions, even the most stoic of us must admit to a fear of future impoverishment. It is this fear that defines our sense of responsibility to others. As a general rule, the further that we are removed from a state of satisfaction with our basic needs, the more insensitive we become to the needs of others, with extraordinary exceptions.

What arises out of our personal circumstances is discrepancy between our personal needs, or more correctly our wants, and the needs of others. An internal conflict of interest arises with respect to when to direct our values to our own accounts and when to direct them in the service of others. We become split personalities. The duality exists in each individual by dint of our sharing common territory. Whether we acknowledge it or not we are interdependent. This moral implication of responsibility translates to a social obligation by virtue of our congregation for the primary purpose of enhancing the security of each individual. Service for the disadvantaged takes the form of the creation of opportunities for others to raise themselves above the meanness of habitual self-service to a higher moral stature through rendering services to others.

The Social Contract

This relation between individuals and their societies has come to be known as the social contract. The social contract presumes that man is freer within society than he would be on his own. Underpinning this assumption is the necessity to examine the natures of freedom and cooperation and only then to accept the premise of the contract. The basic idea of the social contract poses the interests of individuals in opposition to the interests of groups of individuals, a relative mix of which motivates all political activity.

While societies have been constituted through the promulgation of rules for as long as there have been societies, it was not until the seventeenth century that social structures were examined theoretically to formally compare relations between costs and benefits. What underlay this movement was the moral imperative that argued for government as it 'ought to be', distinguished from government as it was then known to have existed. The basis of such theories was a comparison between life for mankind as free individuals operating under the laws of the jungle, versus life as it could be if secured by mutual agreements to collaborate in the pursuit of common interests. We can readily recognize the analogue of the free body in physics, wishing to be independent as exhibited by inertia, and the reality of its relations to other supposedly free bodies.

The emergence of the social contract as the bona fide concept underlining all civilized human relations evolved incrementally. Thomas Hobbes's view of social organization was that, since nature was a state of continuing hostility between individuals, the only conceivable recourse open for enhancing personal security was voluntary subservience to the dictates of a sovereign better able to prevail over adversities in the interests of all under its authority. To accept such a premise necessitated forfeiture of a degree of individual freedom and subscription to implied obligations, as to defend society against external hostility in the collective interest. John Locke augmented this basic premise with the addition of protection of personal property, arguing that personal property enhanced the ability of the individual to maintain security. One consequence of this additional freedom was to tilt the balance of security disproportionately in favour of those with property compared to those without, the effect of which was to generate grounds for resentment that undermined the general interests in security. Questions were inevitably to arise: how much disproportionate security is too much, what are appropriate social controls of abuse and how should the path to optimal advantage be defined? The reader will have recognized that, for the first time, morality surfaces as the arbiter of social order: not 'law' but 'justice', a profoundly different concept. The implications were enormous.

In his treatise, *The Social Contract*, Jean-Jacques Rousseau clearly articulated the minimum requirement necessary for government if the moral high ground was to be maintained: the form that government should take must rest on the consent of the governed.

The first application of this principle was to be no less far reaching, not simply by virtue of its adoption, but for the clarity and daring with which a new nation demonstrated its resolve to hold the principle of 'justice' above the 'law'. The Constitution of the United States of America articulated in law the essential rights and responsibilities of citizenship based upon popular consent, subject to judicial amendment that relied upon the same principle – upon which a more beneficial form of government was to rest.

112 Intention: Orientation Towards Goals

Intentions

Intention suggests a stretching out or straining, a state of being 'in tension' for the purpose of extending beyond the present state or condition. To intend therefore implies work. Such work, where it is applied to the formulation of goals, is mental. Few of us habitually formulate goals on a continuing basis for our continuing futures. It is simply too taxing on our mental stamina to anticipate each new moment and event and prepare for the greatest returns on our investments in time and energy. We rather tend to be selective as to what priorities more obviously, necessarily or desirably require our prior attention, and plan accordingly. The events in between just fall into place within the increments of time that span between.

These are the routines of life that we perform without premeditation or thought, or if thought is required it is superficial. These latter occasions are the instances when we fail to recognize the importance of relations and opportunities, when we tend to become the victims rather than the victors of circumstance. Then there are those who do not appear to anticipate or plan at all. They are the reactionaries amongst us who deal with life's situations as they arise, reacting spontaneously to impending change.

Intentions are the products of a state of mind whereby alternative futures are weighed for the purpose of ascertaining the most desirable. The 'in' refers to in-process, the 'tention' to tension as between competitive candidates for highest priority. Such products represent conditions that are essentially different from those of the present, and potentially more beneficial. In this sense intentions are products of design the processes of which are analogous to a morphological continuum in which the clay of reality is constantly being kneaded in attempts to mould hopes into expectations. The challenge therefore is to identify the forms in which dissimilar forces converge, and the means by which they may possibly achieve resolution. Only then can we consider which particular resolution, or set of resolutions, can best serve the circumstances of a particular time and place, and resolve to apply leverage in support of the most favourable prospects. Thus intentions are preparation for action.

Purposes

Insofar as we are unable to reveal evidence in support of any predetermined purposes to life, any purpose to a particular life can only be defined by the intentions of that life, the intent to remain alive being the most pervasive and compelling. Those who ascribe purpose to some external power do so of their own free will or under the influence of others. In so doing they too act under the auspices of individual intention. Even under duress we act in our perceived best interests, albeit under conditions often narrowly prescribed by circumstances.

The underlying purposes of intentions are therefore self-betterment. For those of us who aspire to better circumstances, we are confronted by a problem of our own making: the definition of 'better'. Not unequivocally better, but better under specific circumstances, which circumstances are constantly changing. Generally, better means better for the first person singular, me. We need to reiterate the question 'what is better?' over and over again as we proceed forward in time in order to assess the nature of our relations to the constancy of change. As with goal formulation, judgements with respect to self-interests may be habitual, intermittent or reactionary.

The roots of egoism, the philosophic belief that good is defined by the pursuit of self-interest (not to be confused with egotism, the overstatement of personal importance), lie in antiquity. History is steeped in expressions of such belief, from Hobbes through Spinoza; goods that Hegel attempted to reconcile through synthesizing the interests of the one with those of the many. Ludwig Wittgenstein, the early twentieth-century philosopher, linguist and semanticist, held the position that, to be meaningful, statements must be verifiable by testing in relation to the physical world; a scientific approach. He later came to modify this view allowing that intentions are what words convey, while he still held to the view that words only made sense if our understanding of them coincided with what they did, that is to say, the way in which they are used. In this way he gave impetus to the idea that words are our best efforts to reconcile our understanding of the world with the way it actually works. A metaphysical word like 'miracle', measured by these standards, does not carry any such connotation of reconciliation between our felt experience and our physical environment and therefore doesn't make sense.

Indeed, the history of the twentieth-century has, to a large degree, been moulded by egoism, often inflected by a disproportionate measure of egotism. The command of mind over matter has progressed with command as the means 'and' the end, or rather the means to the acquisition of greater power to command greater ends, from whence to proceed to positions of greater command. This, as we shall discover, has inevitably led to change in favour of collective interests, the doctrine of the greatest interest of the greatest number, when personal interests come into serious conflict with those of society.

Control is not a civilized end in itself but rather a perpetuation of narrow self-interest unless directed to serve the collective interests of life *in toto.*

Goals

Goals are the objects of intention. Intentions serve as starter markers in pursuit of improvement. It is not, however, sufficient to want to 'go' unless there is a corresponding interest in wanting to 'arrive'. Departures lead to any number of destinations, therefore, before departing we need to mentally ticket ourselves for an arrival at a particular condition of change in circumstance or possession. That destination is our goal, and anticipation of arrival is the motivating force behind our actions.

Priorities are imposed upon us or elected based upon self-interest or compromise with the interests of others. Insofar as intentions are subjective, all intentions may be viewed as being directed towards self-fulfilment. Services rendered for the primary purpose of benefiting others are rewarded with a particularly piquant dividend of self-satisfaction. We have earlier discussed the importance of sequence with respect to facilitating or denying futures. It is priorities and action in support of their implementation that determine sequence and therefore destinations. It is good intentions that make an action good quite independent of advantages that may or may not flow from that act as beneficial results. One should establish goals on the basis of principle, recognizing that means may involve compromise.

A sanguine interpretation of the last paragraph of Voltaire's *Candide*, in which he advocates that we 'cultivate our gardens', is that we should hone our positive attitudes and orient our tasks towards the enhancement of our natural abilities. A more exalted interpretation would recognize that, in cultivating one's garden, one should embrace a goal so lofty that the literal fulfilment of it would be inconceivable. It is not the scaling of summits that affords us enduring rewards, but rather that summits epitomize our ultimate intentions and thereby give our courses direction.

113 Commerce: Materialist Affairs

Needs

Insofar as material needs are the most easily identified, targeted and satisfied, it is to the acquisition of material things that most of us direct our attention much of our time. The process by which we satisfy such needs is through exchange, either directly by bartering or through the giving and acceptance of token equivalence in value.

Specialization

Reciprocity is the founding principle upon which civilization has evolved. Self-sufficiency whereby man answered only to himself for his personal survival and wellbeing died in Mesopotamia 10,000 years ago. The hunter-gatherers divided to become hunters and gatherers. Each depended upon the other for their particular speciality. Man has divided responsibilities for providing for his needs to an ever-increasing degree ever since.

Excesses

As populations have enlarged, the scope of each speciality has been further subdivided. The tradition of the division of labour, formally consecrated by Henry Ford and trade unions, extends to splitting hairs. Today one person dots the 'I's (pronounced eyes) while another crosses the 'T's (tees) to the extent of inducing a state whereby our eyes become crossed and our minds dotty!

Abuses

Coordinating this humble, humdrum and often humiliating work has become the speciality of a cadre of people who exploit the dependence of the many for the benefit of the few. Masquerading under the guise of leadership, tyrants, emperors, kings, lords, popes, generals and, more recently, captains of industry and commerce have asserted claims on the lives of large populations. To divide and conquer has been their unabashed modus operandi.

The System

The idealized principle of commerce is that all transactions should occur between a willing buyer and a willing seller, but commerce is not quite so simple since the interests of the parties are fundamentally opposed. The buyer wishes to pay as little as possible while the seller wishes to receive as much as possible. So, hiding under the mantle of the idealized principle is an inherent tension between the parties that manifests itself in a need to compromise through agreement or coercion to make a deal. However, certain transactions cannot be viewed as strictly or simply willing but arise out of necessities. The need for food, shelter, clothing, medical services, etc. drives the buyer to buy regardless of the asked price if it is within his means. To this extent the buyer becomes a reluctant buyer, particularly if the cost of his necessities places grave financial stress upon him. On the other hand, the seller can generally afford to disregard the prospective buyer of marginal means and target his products to the average consumer or to a particular class of clientele, pricing accordingly. From the buyer's viewpoint it can be seen that below a certain level of privation his options are reduced to reliance upon charity or crime to satisfy his essential needs. From the seller's viewpoint, in order to extract the maximum amount from a buyer, he may elect to exploit certain idiosyncrasies of the marketplace. For example, by reducing the supply (production or competition) the seller can command higher prices. By specializing in the sale of products or services viewed to be critical to the wellbeing of the buyer, e.g. medical, legal or financial services or products, the seller can place a premium price on his offerings understanding that the buyer views these transactions as a matter of urgency.

So we have, simplistically, converted a somewhat naive, idealized free market principle into a crude system of exploitation of human relations for short-term personal gain. The whole business of business has become, if not atheistic, agnostic or godless, unless of course it is the business of selling Bibles. To rectify this, conservative businessmen often use God as a front to give them an aura of respectability through association where they would otherwise be shown wanting.

Trading in Commodities

In the theatre of commerce the small have succumbed to the large. Impersonal corporate structures have relentlessly overwhelmed private enterprise, depriving it of its independence of action. Largeness is their largesse. Multi-nationalism is the mechanism by which they simultaneously expand markets and circumvent laws and tax obligations promulgated by sovereign states. The aggregation of many sources into few in the name of economy of scale has had the effect of subserviating the wishes of the many to the commands of the few. Five hundred people control 50 per cent of the world's wealth. As the marketplace moves closer to becoming monopolistic, the many no longer control patterns of supply and the few suppliers start to raise prices. At the root of this phenomenon is the trend towards aggregation of ownership through stockholdings. The owners no longer have a direct interest in, knowledge of or responsibility for the products, but merely in the profitability of the business. They place the employees of the corporation in much the same subservient position as they place the public, beholden to corporations for their livelihood. The owners are fickle. Corporate structures are viable only as long as they are adequately financed. Such viability does not rest simply upon the quantity, quality and cost of a corporation's products, but upon the relative attractiveness of other corporations to investors who are ready to switch their allegiance and capital from one industry to another without hesitation if a better prospect presents itself. It is a better return on their investments that they seek. The basic premises that formerly underlay commerce have been supplanted by those that define capitalism. The primary commodities (goods and services) are no longer primary. Today's economic life is based upon trading in token equivalence in values where commodities are merely identification symbols.

Trading in Futures

Invoking our natural interest in projecting security further into our futures, that interest is now being applied to the projection of economic security into the future through the device of 'fixing' future prices. This practice, known as 'trading in futures', involves contracting for the sale or delivery of stocks or commodities at fixed prices for delivery at agreed future dates. While security of knowledge of future costs passes to the buyer of such 'futures', there is no security for consumers that a proportionate advantage will flow to the public. Thus, trading in 'futures' for the benefit of the few may well translate to the trading-in of future consumer advantages due to prior price fixing.

Options

In a truly open market the many may buy from the many in the absence of deceit. That, it may be held, is truly democratic. A multiplicity of sellers assures competitive prices and choice. By contrast a limited market manipulates prices and restricts choice. At issue is not which is the better system per se, but what form of organization of labour and capital can better generate the means by which to produce affordable goods in quantities of a quality to satisfy the demands of a burgeoning population without abuse and negative consequences. It is weight of numbers of people that safeguards public interests in a democracy. Conventional wisdom would therefore suggest that some form of cooperative ownership in which the stakeholders (the consumers) are also the stockholders (the sellers) would be the most beneficial and just system for the greater number of people. Extending that idea to its logical conclusion, the state (representing all the people) is the obvious candidate to implement such a reciprocal programme. That too, it could be argued, is truly democratic. Unfortunately, such a solution negates two factors essential to viable commerce: competitive pricing and personal incentives to make the system work.

Some kind of hybrid mix of scale and multiplicity is implicit in any future viable commercial format serving large populations wherein there is tangible popular interest in the enterprise while preserving the incentive to succeed through maintaining competition.

The Shades of Business

One of the disheartening things about the business of business for a young idealist is that, regardless of the nature of a business (to which one is presumed to have been attracted in the first instance for its intrinsic values), the highest priority appears to be directed from creating to maintaining and subsequently towards exploiting the enterprise.

Trends

The age of titans is emerging whereby a single military superpower nation, multi-national corporations, monopolistic industries, monolithic skyscrapers and monosodium glutamate in all our processed foods, propel us rapidly towards a monoculture by default.

The increasing incidence of appeals by private corporate enterprises for massive government grants to forestall bankruptcy requires rethinking. The competitive system is driving wealth development into the hands of the incorrigible few. Burgeoning social interest in the detrimental effects caused by narrow interests is going to accelerate the cause and course of unity of social purpose. Corporations will slowly come to understand that antisocial behaviour will not be tolerated or permitted. The trend towards the consolidation of corporate wealth at the expense of diversity is the greatest assurance that unity of purpose will prevail, though the form of that unity and the purposes served will be unrecognizable to those who are promoting unity today in the pecuniary interest of stockholders.

Excessive democracy is anathema to the business world because it poses the prospect of ultimate authority of the popular will over otherwise unmitigated commerce. The strategy of commercial interests has therefore been to coerce or otherwise deceive public opinion into believing that the ultimate freedom is one that is not held ransom to political will.

In a world under democracy, nominal control is vested in the people who, being spared corruption and catastrophes, will ensure that the meek shall inherit the earth.

Caveat Emptor

What is fundamental to restoring dignity to human relations in general and commerce in particular is a need to place limits on self-serving practices for the benefit of society as a whole, of which there are but two means. The first is the exercise of moral or ethical precepts that place voluntary curbs upon exploitative commercial behaviour. These are the means by which the average decent human being behaves in society. The second is for society to rein in those who exceed the limits of generally acceptable social behaviour through the introduction and enforcement of rules, including taxation. While this may temper most cases of abuse of privilege, there is one class of society, the legal profession, that by dint of their specialized knowledge dominates the legislative, executive and judicial branches of government, defining and interpreting the rules in its own interest. We shall revisit this subject.

Dignity, that quality of confidence in the belief that each one of us is important in his own right, depends upon the attainment of a median balance between self-service and service to others that is the crux of civilized life. To a large degree the manner by which we allow commerce to evolve will confirm or dictate how balanced we are as individuals and as cultures. The open question still stands: what should 'free trade' be free of? Buyer beware!

114 Design: Planning Futures

The hypothesis underlying all planning and design activity is that time, space, energy, matter and relations can be orchestrated through the instrumentation of the intellect, volition and the affections in the interest of optimizing beneficial futures. Planning is the process whereby we attempt to foreclose disagreeable events in advance of their occurrence and to foreopen agreeable alternatives.

Almost all questions concerning design may be reduced to a single question: what is the most significant relation between matter and energy at a particular place and time? Almost all answers reduce to the most fitting or agreeable balance between reason and the emotions as determined by volition.

Time Framing

To plan the use of time is to calibrate the middle of an infinite time line into parts that coincide with the duration of past and intended future activities. To do this we need to understand the historical forces that have enabled us to progress to our present condition. We are 'surfing' on the 'now' wave of time. We have cruised and been bruised in the past, but ahead loom changing conditions. We need to project where the present forces will lead us and assess whether we can intercede in some way to enhance our prospects of and confidence in positive futures. While we 'surf' we 'fish' for better, more defensible or less painful courses. We weigh our options and weight our lines to bias our decisions based upon desirability and probability. We have developed a new science to that end, variously called Operations Research, Decision Theory or Risk Management, which unfolds optional futures before us in the form of decision-trees. Their branches are lines of reasoning, their leaves probabilities and their fruit expected values.

Spacing Out

'Where' we want to be is spatial. Where we 'want' to be is special. In allocating place and space we disclose our values. We may choose to move ourselves, as from place to place; or rearrange our space to meet our changing needs and desires. How we make such decisions is a process of reconciling reason with the affections. Space planning is the profession and practice of faith in our abilities to shape our future use of space.

Powering Up

What is needed is whatever it takes! If we do not have the requisite means with which to achieve our goals, no amount of desire or wishing will deliver them. Thus, the adequate allocation of resources becomes the motive force behind any plan of action. Planning by definition cannot be construed to be an abstract exercise. Rather, it is an exercise in abstracting and assigning resources in support of the pursuit of goals. Economics echoes!

Stuff

Matter is the stuff of common experience verified by common sense. Things are the substance of substantial improvement. The accrual of things is a sign of action that enables further action. The accrual of token things represented by money provides fuel for future action in an as yet indeterminate form. Thus, the organization of things becomes central to the planning process. Planning the intangibles of time, space and energy, and the tangible characteristics of matter, effects the efficient supply, distribution and utilization of these quantities in order to achieve higher performance and thereby gain greater benefits.

Knitting

The effect of time framing, spacing out, powering up and stuff is to empower dynamic changes in relations. None of the above can be achieved without all of the above. How we effect change depends upon what we do and in what sequence. One plain plus two purl stitches does not a sweater make. Perseverance is required to form the fabric of our goals through the cultivation and nurturing of correct relations – one stitch at a time. In planning we plot economical courses to those ends as a matter of principle.

Intelligens

What we do and in what sequence are functions of the intellect. The intellect is the manager of knowledge. To the extent that we plot our courses intelligently, we enhance our probability of success. Thus, intellect, the capacity to understand correct relations, is our navigational aid in making arrangements that best serve our greatest interests.

Volition is Volatile

Volition is the arbiter between our heads and our hearts. As such we are susceptible to change of mind as to whether reason or the emotions shall govern, and to what degree. Our decisions are linkages between events, flexible in planning futures, torpid in the present and rigid in the past.

Affectionately Yours

Perception, that sixth sense that intuits comprehension, is self-determinant. Through such feelings we make connections that distinguish good from bad experience. What we care for most defines our values, which determine our goals, which frame our decisions, which impel our actions. Finally, we enlist reason to justify our actions and reinforce our affections.

By Design

Much as planning defines goals, allocates resources and facilitates action, the design process imbues our plans with qualitative aspects that are directed through reason-based decisions in the interest of securing fitting relations. Thus, design may be viewed as the process of searching for and establishing fit, fitness or fittingness with regard to the relations of separately identifiable parts. If all one achieves in design is to create something interesting, then all one has done is to stimulate interest in others to do what the first designer should have achieved in the first place. In design we attach significance to the manner in which parts fit together to form totalities.

Design draws connections between ideas and actions through contemplation, intention and devotion, outlining the disposition of parts proportionately. The plan addresses general strategic issues, while the design defines objectives. However, the word 'design' connotes a process and a product. Knowledge of one can be used to help reveal the nature of the other. The end is the fulfilment of the beginning. One can only design something of which one has a vision. 'By design' signifies 'by intent'.

To search for an ideal plan is an eminently defensible, legitimate design strategy, only to be compromised by the designer as environmental circumstances dictate, and then only in accordance with design priorities. Logically, the approach should comprise multiple ideal plans, each optimizing performance from the standpoint of a single criterion, e.g. for function, structure, economy, aesthetics, etc., later to be reconciled. However, one seldom has the luxury of adequate time and resources to develop such ideas fully, except perhaps in the design of the simplest objects.

In design temporal problems are manifold. Specifically, and most illogically, the designer is expected, indeed he is often under contract wherein time is of the essence, to conceive of and tangibly represent in resplendent detail, something that not only has no precedent but has no generally acceptable definition: a future. To plan or design demands the definition of appropriate futures. Such a suggestion raises serious questions as to, for example, whose interests are to be served, those of individuals, those with vested interests, the interests of society generally, or some amorphous mix? For better or worse, significant change can be effected by individuals or small numbers of people without the sanction of, or opportunity for intervention by, the great majority.

One thing that history has taught us is that knowledge is conditional. It is based upon times, places, people and circumstances that are subject to change. Understanding this, we can interpret that all current levels of knowledge in all fields are already history, not in the utilitarian sense of obsolescence, but in the definitive view. This being so, our only intelligent recourse is to project ourselves forward, that is, to make change occur that optimizes our greatest future advantage.

Design has something to do with propriety, thereby evoking allusions to integrity, if not morality.

Process

Definitions of sound planning and design practices as they pertain to public interest (and that is what public officials and institutions are pledged to promote) appear to hinge upon the clear separation of personal prejudices from community needs just as surely as we today recognize the necessity to separate the domains of church and state.

A planner's task is to facilitate better futures, not to barricade against change. In the public sector this translates to mitigating need in priority over rewarding greed. To this end the less fortunate merit greater aid than the fortunate. We don't fill hospital beds with healthy people, give educational opportunities to those that already have them, or offer free shelter and meals to those that can afford them.

Planning deals with defining goals and the assessment of ways and means of achieving them. We tend to be strong on the former and weak on the latter. This is not too surprising for it is with means rather than with ends that we generally disagree.

Design is the process of synthesizing abstract ideas and correlating physical elements in order to gain valuable insights into beneficial relations. We need here to make a clear distinction between personal integrity and the integrity of ideas that emanate from an individual. A clear distinction can be seen between the two in examining the personal character and professional contributions of the American architect Frank Lloyd Wright. Identifying a more pompous, egocentric and arrogant personality in the field of professional planning and design would be a challenge for anyone; yet within this same person resided a masterly capacity for synthesizing the integration of diverse spatial, physical and aesthetic experiences of a truly novel and valuable quality.

Creative activity performs a profoundly fundamental function: changing society for the good. In recognition that all that extends ahead are uncertain futures, all that we can do to enhance those futures is to plan by design. To embrace such an agenda we need a clear understanding of needs.

The Nature of Design Decisions

The processes of planning and design involve searching for fitness in relations. Such fitness is characterized as an absence of misfit. It is so because misfits or conflicts are more conspicuous than the absence of conflict, which has the quality of being transparent.

When we engage in design we seek to describe a condition wherein all conflicts are resolved. Such a description is the specification of a fitting solution. In order to achieve such an end we need to ask ourselves with what does such a fitting form necessarily require to conform? The answer resides in its context.

The most obvious needs are physical. Round pegs don't fit in square holes. So it is to physical relations in a tangible world that we generally direct our attention. We cannot make our designs work if they are in physical conflict with their immediate surroundings. The same principle applies with regard to the interests of economic, socio-political and aesthetic fitness.

Therefore, in order that we may proceed with a design process, we necessarily need to define such parameters that are categorically required or desired to be addressed, and then to prescribe acceptable limits within these terms of reference. Such limitations are generally defined in negative terms for reasons earlier described, and take the form of codification for general application.

Now the designer has an array of references against which he or she may test intuitive ideas with respect to whether they pass or fail specific tests of fitness. Trial and error may then proceed from this point in search of the best fit until available resources of time and money have been expended.

There remains the question as to who makes or should make planning and design decisions and under what circumstances. In a rational world we determine this on the basis of criticality. Truth and consequences kick in. If no socially detrimental consequences arise from allowing a person to (say) define the design parameters of his home, then no social justification exists for interfering with his agenda. Suffice to say, under otherwise benign circumstances, those with more at stake are generally better qualified to judge and act. However, in a complex and dynamically changing world, the fact of change may change the action required and thereby the qualifications and identities of those most competent to act. Under such conditions the fallback position has been to share responsibilities for decision-making.

To justify the need for collateral judgements one of two conditions appears necessary: first, where complexity is so evident and pervasive that it warrants the expertise of specialized authorities; and, second, where decisions involve the legitimate health, safety and welfare interests of the community.

In the first case there is generally little difficulty identifying the disciplines that bear upon the issue and introducing them into the process of decision-making. In the second case, specialized expertise is often trumped by political will and authority. The consequences of decisions made by those unfamiliar with, and generally disinterested in, the details of the subject range from trivial to catastrophic.

The propriety of opting for one form of decision-making rather than another generally depends upon the scale of the undertaking. Who the arbiters are and how arbitrary their judgements are in the context of the decisions to be made determine how appropriate or otherwise their actions are.

Synthesis

The process whereby a general thesis or idea of what conditions should prevail is put into acceptable relations with existing conditions is synthetic. To synthesize is to empathize with, synchronize, harmonize, capitalize and ultimately formalize an indefinite number of elements into what has hitherto been just a loosely defined concept.

The significance of complexity and the entropic principle become more evident with an expanding body of historical records that expound upon the results of trial and error under changing circumstances. Insofar as there are almost limitless numbers of criteria that could theoretically be brought to bear upon a decision, and a limited but not absolute limit to the capacity of man to address each criterion objectively, we confront a need to self-limit their number. Hence Occam's razor, to which we earlier referred.

From the foregoing, necessity dictates that criteria be ranked in priority in order to get the components of a series of decisions in the most advantageous sequence, understanding that, as we also discovered earlier, sequence dictates and denies possible outcomes.

Understanding that less complexity facilitates greater clarity and comprehension; with fewer elements a higher probability arises that they can be brought into optimal relations to achieve intended goals. The design process becomes more manageable by limiting the number of factors bearing upon any single decision through piecing a complex design together progressively. Mating simple parts to form simple mechanisms, and successively mating these semi-autonomous subsystems into more complex organizations, enables the construction of compound structures that exhibit a degree of complexity inconceivable at the outset.

Returning to the notion of priorities necessary to achieve fit, we can represent optimal fit as that condition that is free from conflict; one in which all forces are reconciled proportionate to their relative potential to contribute to a fitting design solution. Theoretically these may be represented diagrammatically by force-fields that express how each element of a complex pattern is constrained by those that bind it, placing natural limitations upon the extent to which each influences all others. While we are free to shape form as we see fit, we do so understanding or not that the context within which we work places limits upon the ultimate form that a design takes. In better understanding the context we gain a negative impression, as a cast in the mind, of what forms a design can and cannot take.

If a designer fails to fully take into account the larger context of a complex undertaking in which he is engaged he cannot be faulted by the fact or act itself, but only admonished for lack of a more comprehensive vision. In fairness, finding fault predicated on failure to perform what most people consider practically impossible anyway is questionably justifiable!

In more clearly understanding what is not possible, the designer is guided towards what is possible, from which position he can conceive comparative models and test them for performance. To the extent that we cannot fully understand the forces at work that constitute the context, our prospects of finding a fitting solution diminish. The persevering designer retains faith in his conviction that the longer and more diligently he searches for fit, the more likely he is to find a better solution; or, to put it in its proper context, to find acceptable resolutions to the conflicts imposed by the elements of the problem and its environs.

Ideally, the designer starts with a summary description or programme listing the most significant constraints to and expectations of a design process. In this way the range of possible negative impressions is limited and others are afforded prior notice of the principle values attached to the exercise before commencement of the design work.

Of these programmatic statements perhaps those most demanding are the expectations of what the completed design is intended to do. Functionality is often the ultimate design intention, recognizing that the ultimate measurable performance may be governed by the inherent or extrinsic constraints that attend the design process.

Understanding the above rationale, clichés like 'form follows function' reveal a measure of truth while concealing the more complex nature of design, and, through over-emphasis, on occasion mislead designers in their search for an optimal balance between competitive interests impacting an appropriate design solution.

Design Economy

Finally, let us look briefly at the effect of initial cost constraints upon design decisions, and conversely, the economic consequences that devolve from designs where no account has been taken of the collateral consequences attributable to designs that are governed by such constraints.

A correlation between the impact of severely constrained initial budgets and long-term cost consequences can be shown in the design of a typical building. On the one hand, limited resources severely restrict the quantity and quality of labour and materials used. Reduced quantities limit ultimate performance while limited quality detrimentally affects durability. The consequences of such shortcomings have been demonstrated to reveal that initial cost reductions purported to be 'savings' translate to additional long-term cost expenditures for maintenance, repair and replacement during the useful life of a building of an order of magnitude of approximately twenty times initial costs!

The second set of consequences, less obvious than the first, is the effect of causing imbalances between demands and replenishable supplies. These are not merely allusions to over-harvesting limited resources, serious as they are, but to the masked consequences of negative impacts upon health, welfare and long-term survival prospects of human, animal and plant populations alike. Such considerations transform the formulation of design criteria from the status of being of local interest and subject to expedient decision-making, to worldwide relevance. Such is the nature of and need for holism in our systems of decision-making.

In 'Nature' we behold the supreme designer; in her magic the meaning of life. Her miracles lend cover to the laws that she is simply not ready to reveal. God is a six-letter word! The challenge is not to emulate her performance, for that would be too humbling, but to secure in our designs a sensitivity to nature through association in such a way that people will understand that design in service to nature is of itself in service to mankind.

While the means to these ends are ellusive, the sources of inspiration are infinite.

115 Needs: Exigencies of Necessity

Needs are accoutrements upon which living systems depend for their wellbeing. Being well is a relative term. At the most basic level absolute needs are those properties that are absolutely necessary in order to qualify a system as being viable. For the living, viability depends upon enduring opportunities to participate in the principle functions that attenuate life; namely formation, sustenance, development, vitality, procreation and, by way of a legacy to future lives, mortality, to all of which we have earlier referred. The world of the non-living has no needs and no worries.

Necessities

In order to proceed we need to be assured continuity in our existence in the most elementary ways. Our most urgent desire is to remain alive. To subsist, we require water, food, security and protection from inclement weather: a minimal survival kit. This was the basis upon which the Babylonian settlements engaged in ranching in preference to ranging. Yet at the dawn of the twenty-first century A.D. we still have not achieved the most fundamental of goals: a sustainable human population. It is not that we cannot achieve approximate equilibrium in our natural setting, we simply don't have the collective will to do so.

Luxuries

For those of us that attain the ranks of survivors, we continue in our pursuit of personal gain. Why? Because we feel better and more secure when we are better provided for. We enter a class that generates demands for more diverse, more useful and therefore more advantageous commodities, none of which is essential but all of which, in our individual judgements, give us additional satisfaction. We become discretionary consumers.

'How much is too much?' we might ask. The objective answer is: when we exhaust resources faster than they can be replenished. The subjective answer is: when we incur the wrath of less fortunate people to the extent that they take action to secure a greater share of the benefits. Then we know that we have crossed the line between the benign and the asinine.

These are the thresholds of waste and greed beyond which we are unable to excuse or morally defend our actions. Needs to survive morally pre-empt discretionary needs. The survival kit bag carries a higher priority than excess baggage. The test of this rule is to put oneself in the shoes of others. To breach this caveat is to invite the breakdown of civility, and ultimately of civilization.

Advances Beget Advances

Every advance that enables a new benefit sets the threshold of expectations a notch higher in the continuing search for future benefits. The effect of such upgrading is to generate new and unprecedented demands, the attainment of which places a greater strain upon finite resources and on human tolerance for disparity between the quality of life for those that attain such benefits and those that do not. Such distinctions engender presumptions of entitlement in those that enjoy greater benefits that prompt them to defend their privileged status. The less fortunate generate deep and well-founded resentment on the grounds that precious resources are increasingly being squandered by the privileged in the pursuit of frivolous pleasures.

Social Orientation Towards Goals

In order to counteract the tendency towards fragmentation and a breakdown of society, benefits require to be more generally distributed. This is the process that underlies the advocacy of human rights upon which the social contract is based and civilization hinges.

Human needs underpin claims to human rights. Two authorities legitimate rights: morality and law. They coexist in a continuous state of disconsonance; morality preceding law in its advocacy for socially beneficial improvements, while changes in law attempt to confer reconciliation. Claims to human rights rest on two fundamental predicates: freedom from want of necessities and unanimity in the accordance of human dignity. In order to maintain stability in a democratic society we necessarily need to distribute its opportunities more equitably. This, as we have ordained in our constitutions, is the crux of our self-chosen course in the pursuit of human happiness. How a society assesses and fulfils these needs is the measure by which it is ultimately to be judged and held accountable.

The key to understanding this exigency hinges on the development and maintenance of a judicious sense of proportionate relations between one's own needs and those of others; what we might think of as pre-Einsteinian relativity.

116 Expedience: The Pragmatic Option

As we discussed the subject of principle we recognized the general rule that the operation of principle served to enhance probability. We now entertain the alternative question: under what circumstances shall we make exception to the general rule of principle and, in effect, advance special rules that override general rules?

The distinction of the general rule is that it is generally distinguishable, from whence it is a short step in reason to hold that it should or shall be generally observed as a matter of law. Laws in general, and particularly in a democracy, tend to work in the interests of most people most of the time. But what happens when rules conflict with self-interest? If they are not governed by law, such principles are not merely open to question, but open to disregard. There is an aphorism in the

Balkans that likens law to a closed gate in the middle of an open field. So the rule of self-interest would suggest that, where interest conflicts with rules, one should follow the rule of nature and the line of least resistance.

Expedience

Expedience is the shortest path between decision and expedition. When we make decisions our self-interest is best served by putting it into effect by the most economical means, that is, to expedite it quickly. So expedience becomes the preferred path in matters of self-interest up to the point where to break with principle (rules) renders the expedient path no longer the most economical means to an end. What we are unravelling here is a general exception to the general rule that states that it is OK to pursue any end by any means just as long as one doesn't contravene the principle of economy of means. This is another way of stating that it is OK to break the rules, or even the law, if one doesn't get caught and have to pay the consequences. We are led by this stream of argument back to the act of decision-making whereby we must attach 'flavour' to our decisions that guides us to the sweetest of possible consequences. That 'flavour' is morality.

Pragmatism

Pragmatism relies upon the doctrinaire belief in practical and probable consequences for its authenticity.

Options

There appear to be two kinds of decisions: those taken on principle that are defensible by the intrinsic logic of cause and effect, and those taken pragmatically that resort to the extrinsic logic where self-interest becomes the overriding determinant as to how decisions should be made. The former suggests that accountability is the better, the more responsible part of valour; the latter that expedience is admission to not knowing or caring enough to know better.

Decision-making seems to boil down to the opposition of principle to pragmatism, either one of which can get one into deep water fast.

Nature acts expediently as a matter of principle, and vice versa.

117 Concern: Relative Interest

The word 'concern' is simply a synonym for the idea of care as self-interest. If we argue that this is not so, insisting that the term as we apply it be directed towards the welfare of others, we cannot disallow the follow-up question: does this hold if we are unaware of their circumstances? We are either concerned about the conditions of others because we are aware of them, or we are concerned because we are not aware of them. Insofar as both conditions are functions of the state of our knowledge as it relates to our values, that is what is of primary concern to us. The condition of others may or may not be the driver.

Subjective concern is primary because it is central to both states of being aware and of being unaware. The object of our concern is secondary because, while in the case of our knowing of a circumstance our concern may be warranted, in the case where we are unaware of a condition we have no grounds for being concerned.

So when someone states 'I am concerned …' we should first enquire what the state of that person may be (for which we may have our own grounds for concern), and only secondarily what the ostensible subject of their concern is.

Concern is a relative term. Under circumstances where there is little margin for error, the meaning of concern becomes clear: what is judged relevant is included, what is judged not is not. That is what matters until errors or omissions in judgement come to our attention, at which point we redefine relevance and run through the exercise again. In so doing we prove that concern is relative to knowledge and then interest; our knowledge and our interest.

In order to achieve anything of significant value it is necessary to attach exceptional importance to that achievement. Without such concern, one may enjoy the process and be satisfied with the product of one's efforts to a degree, but this does not demonstrate anything exceptional – except perhaps the significance of one's ability to represent oneself as ordinary!

We have linked concern with caring. Who cares? You do if you have concern. So concern is relative interest, most clearly distinguishable in the first person singular. Beyond, care is evident in the second person, then progressively to third persons, animals, distress in plants and stress in inorganic matter, all in accordance with a natural hierarchy of priorities that, when they become extended, frequently become the subject of concern extended in others.

118 Utility: Accordance of Value

Value is not an intrinsic attribute of things but rather a product of perception; a subjective quality conferred upon things. In the absence of such perception things have no value, unless we attach value to things and their interactions for being essential to interdependence.

A wheel is not essential to a machine per se, but of value to those who perceive the potential of the one for enhancing the performance of the other. In the absence of running water a bathtub may be perceived by an Eskimo as a place to store fuel for the winter. Insofar as each person holds unique perceptions, degrees of usefulness cannot be measured in absolute terms on a calibrated scale. Nor can relative values of things be measured between people except as they may agree for their mutual convenience. Usefulness does not exist as a matter of fact, only as a state of mind, or in agreement between minds, subject to specific times and places.

Serviceability manifesting itself in relations between persons and things generally bears connotations of value or cost. Utility in this broad philosophic sense may be identified as latent power potentially applicable to good or evil purposes.

A narrower definition of utility, as used in economics, ascribes values to goods and services. In this case desire is the activating force. The problem then becomes one of assessing degrees, in the first case of desire, in the second of value, their not being synonymous. The problems of objectively measuring degrees of subjectivity are formidable. Even ordinal measurement, the placing of desires in a hierarchical order, is subject to changes of mind. Attempts to calibrate subjective degrees of desire render inordinate degrees of imprecision.

An even narrower construct is that purported to represent the social value of a utility. This is a simplistic reduction of values to market economics represented by currency. In the broadest sense dollars and cents don't figure. What is the utility value of fresh air, love or God?

119 Fitness: A Sense of Appropriateness

We have noted the mix-and-match phenomenon that results from the turbulence of change and exchange. Mismatch is the entropic effect, the disparity that ensures that the merry-go-round continues. Match is the brake that tempers the whirligig, but only temporarily.

When we examine what we mean by match we discover to our consternation that the idea is not so obviously the antithesis of mismatch. We use the term match to represent the pairing of like with like, and like with unlike. So when is a match a mismatch?

Mismatching

A standard test of the psychologist of yesteryear was to present a small child with the problem of fitting geometrically shaped coloured pegs into holes cut into a board. The holes had profiles that corresponded to the sections of the pegs. Each peg fitted a single hole. What was a smart child to think of the meaning of the concept match when in her understanding, to match colours meant to place like with like? Now she could establish no likeness between pegs of differing colours and shapes, or three-dimensional pegs with two-dimensional openings. There were obvious mismatches in colours and forms. Was she to look for a new and particular coincidence of expression befitting specific mismatches? Ask that of a small child!

Symmetrical Matching

Raindrops running down a windowpane flow into conjunction along a single path. They combine their identities and couple their energy, accelerating downward in surrender to the forces of gravity. The effect of their combination is not of resistance, but rather of mutual submission. They pool their commonalities of composition, form, energy and direction to a single effect: the neutralization of conflicts of difference. Matching of like to like is abundantly evident in our environment; a kind of herd behaviour in inanimate objects, as dust to dust, etc.

Asymmetrical Matching

What the small child may not have understood was the subtle distinction between matching and mating. Mating is the reciprocal act of matching whereby dissimilar entities are reconciled through counterpoise, offsetting discrepancies of difference with equal and opposing, neutralizing forces.

The positive pegs and negative holes are such matings – circle to circle, square to square, triangle to triangle, etc. – as are magnetic attractions, chemical affinities, sexual pairings and the physics of force and counter-force generally – complements of the universe!

Harmonic Accord

What both matching and mating achieve is a form of union whereby each component of the union achieves compatibility with its partner. Union is to the benefit or good fit of each partner to the other to the extent that it dispels tension between them as accommodation is distinguished from alienation.

All elements, forms and forces tend to coalesce into pools of common denomination. All dissimilar entities find in mating relief from their search to escape from being misfits. Fitness, correctness, rightness and goodness equate. In each case matching is appropriate to needs and an innate sense of fit.

The Path to Virtue

Of all the qualities that propel a person in the direction of leading a virtuous life – sensitivity, tolerance, humility, love, charity, justice, conscience, integrity, commitment, duty and hard work –

only one, hard work, is acknowledged to be profitable in our so-called developed cultures. Little wonder that we do not strive to emulate the virtuous; little wonder that we do not emphasize virtue as a lifestyle; little wonder that virtuous people do not rise to the top positions in politics, business and industry. We have cultivated contemporary cultures that are in effect devoid of values; save the value of money that we have determined is value-neutral, delivering costs and benefits indiscriminately. There is nothing wrong with money per se, it is a necessary and convenient intermediary by which to exchange goods and services, but if we are betting on improving the quality of our lives we are using the wrong currency.

Improvements in the quantity and quality of food and accommodation are recognizable by pigs and goats, so what is it that distinguishes the human being from the rest of the animal world? It is a value system, or rather the ability to redefine or switch on or off our value systems as expedience dictates.

120 Value: Degree of Worth

Value in Hindsight

Plato held value to be the highest and most difficult of ideas. Aristotle similarly held the quality of value to be higher than all others, believing that goodness was an attribute of things, and therefore essential to their being what they are, implying that value was an objective reality.

This classical connection between a thing and its value held until science imposed its standards of testing empirical truth on the eighteenth century. A stream of British philosophical thought, logical positivism and its heir-apparent analytical philosophy, took up the challenge of applying scientific scrutiny to values as evinced in the works of Thomas Hobbes, John Locke and, most notably, David Hume.

Hobbes's take on value was that value itself rested upon higher predicates, peace and security, and that in the absence of peace, values were of moot worth. The logic behind this position implied that all civilized societies hinge upon the maintenance of peace and that all persons require to be secure as a precondition to being civilized.

Arguing that both peace and security may be found in a prison cell, Locke inclined his position regarding values towards making them contingent upon addressing the more general problem of human understanding. Hume was to convert this stance to an inductive science of human nature of which the subject of values was but a small part; premised on the belief that mankind in general founds its highest values upon sensitive emotions rather than upon reason. He maintained that awareness, including that of values, originated in experience derived either from sensation of external stimuli or from the products of sensation associated internally and represented as ideas. As such, such notions as metaphysical values made no sense because they were not so derived.

While what we popularly call reality may be of this nature, what (other) men think, say and do does not always seem to fit this mould. Here we should acknowledge that making sense is not the key to having thoughts and taking actions, but rather that impulsive events analogous to spontaneous combustion appear to convert nothing into something in our consciousness.

Sense, meaning coherence, only transpires when we are able to construct plausible connections between data. Thus, Hume continued, the focus of our awareness is either objective by reason of being accessible to our senses as properties of matter, or subjective as sets of related ideas. This habit of connection, as between stimulus and memory or stimulus and idea, or idea and idea, is all that reason can reveal as being factual with respect to assertions of causation of one by the other. Thus, we believe because we feel that the products of these habitual connections are plausible, the way things are, because they appear to exist and function that way, quite distinct from pretensions to justify them through reason with which there is no necessary connection.

The 19th-century philosopher Rudolf Lotze contributed more than any other to the popular use of the term 'value' through implanting value as the heart of philosophic enquiry. Lotze was compelled by a desire to reconcile the rational backbone of science with the aesthetic life force of art, to link the unity of each with the other under the purview of the natural sciences.

In the early twentieth century Bertrand Russell and Ludwig Wittgenstein were to extend the legacies of Locke and Hume. Russell claimed that the defence of verity in values depended upon the extent to which they were defensible in logical analysis, while Wittgenstein defined values as conditional pictures of reality due to limitations imposed by the normal use of language.

In the latter half of the twentieth century the word 'value' became devalued, being used interchangeably with cost to become not only the centrepiece of contemporary Western life but a cultural obsession evidenced by the degradation of the US judicial system to a legalistic commercial enterprise, its bottom line 'the' bottom line, where the pursuit of justice was no longer held to be the highest virtue. Cost had become the lowest common denominator of almost all private and public transactions where value had formerly been the highest common multiple.

Values as Reflections of Self

Underlying our difficulties with concepts of value are our even less clear concepts of goodness. Feelings, thoughts, will, aesthetics and religion rely upon differing interpretations of the word good as befit their own class of experience.

Personal values may be expressed and understood as belonging to a broad range of classes of goodness from the singularly objective to subjective expressions of attitude. Some confusion is enjoyed by those who attribute intrinsic goodness to natural causes because such good is ultimately traceable to subjective experience, as, it has been argued, is objective goodness.

A jackrabbit may sense that he is the centre of goodness in the world and he is as correct as if the first person singular senses that he or she is. My goodness! Jack is the author of his own subjective values. He is a self-authenticator. The notion of intrinsicality of goodness without reference to authority is, by definition, unauthenticated. When authenticated, the notion is no longer intrinsic. The flaw in belief in intrinsic goodness may be found in the concept that something is worth having, pursuing or doing for its own sake distinguished from a means to some other end. For example, engagement in art, science or gardening may be to purposeful ends but such pursuits are not devoid of personal satisfaction.

What is significant about one's values is not whether they are incontrovertibly sound or high-minded, but that they are the drivers of what constitutes one's psychic id, that subconscious responder to stimuli on a pleasure or pain level. As such they truthfully represent who you are at any particular time and place. We are all what we are for very good reasons, for which we bear less responsibility as individuals than is often assumed.

As to which values are of a higher order, we tend to revert to the self-fulfilment of our prejudices and make further value judgements.

Collective Values

Values are accepted deferentially by those who hold them to be truisms. For these people the boundaries of acceptable behaviour are transparently clear; their values are beyond question and inaccessible to reason. Adherents of such moral codes view and often chastise transgressors as heretics. Such values are fear-driven, such fears value-driven; fear that logic will somehow undermine the security and therefore the value of their beliefs. Then there are those who view values circumspectly, whose values are conditioned by their circumstances, who are pragmatic in outlook and opportunist by disposition. For these, values are personal tools, useful to lever their way through life.

But neither of these views of values allows enough latitude to embrace a general rule as to what values are or do. Values are not rules. If, as is generally held, values are deeply rooted in truth, and if truth, as we earlier have discussed, is fact-based – and if, further, truth is accepted as subjective, relative and circumstantial – this will account for the broad variation in values held. It begins to explain why the epicentres of the great religions are geographically separate, why their creeds are customized to match their experience, indeed why one man may hold that his values are inviolate while another regards his as postures of convenience. Values, generally speaking, are learned. If learned, then to a large degree taught. If taught then by different teachers at different locations at different times. The lessons are taught in different languages and in differing word sequences. Values, we must conclude, are inconstant, changing with the times, tailored to fit the best interests of society for the purpose of protecting and stabilizing the populace. Like truth, values are manufactured.

What are, in general terms, the positions from which values are argued? They are the voices of conservatism striving to hold the course of events on the one hand, and the voices of advocacy for improvement on the other. And of course the definitions of improvement are the battlegrounds whereon the subjectivity of values is arrayed; where the arsenals of those qualities known to be good by experience are ranged against those that have potential for greater good in the future. In a nutshell: satisfaction versus challenge.

What fills the place of values where they are absent? For the individual, values define priorities. One acts in accordance with the dictates of one's values to achieve goals. Values govern actions. In the absence of values momentary expedience dictates actions, individual egos compete without rules and anarchy is the general state of human relations. In societies absent values the most ruthless egos prevail, social order is dictated by tyrants and totalitarianism flourishes. A social code that does not accommodate the needs of man in the singular and in the plural is worthless. But men are fallible. They must be organized in the singular and the plural to correct their errors, their mistakes being as crucial to progress as their verities. Is it not a great blessing that we are so, so that we have cause to work in the direction of improvement, motivated by hope?

Values as Utilities

Values, it appears, are manufactured for the purpose of serving interests, those of individuals and groups of individuals. Like any other artifice they are not intrinsically good, they are more correctly good for some purpose or purposes, which purpose or purposes are not intrinsically good either. If one begins to apply scientific standards of objective scrutiny to values it quickly becomes apparent that values are instruments of relative power. Just as the lever and pulley enhanced man's physical power over that attainable through the use of his muscles, so values elevate a man's moral authority above that attainable without values. So, apparently, what values do is assist the promotion of agendas without regard to virtue, yet it is always virtue that is invoked to defend the stature of values. Ironically, the connections drawn between values and virtues give impetus inferentially to what ought to be, when as often as not where public interest is the measure, such values ought not to be embraced. One must therefore conclude that linkages between values and virtues are suspect. The scientific method, it would seem, turns stones disinterestedly, except in the interest of searching for truth, and under some stones may be found that which some would wish to remain covered from the public view. Little wonder that values are often held to be sacrosanct and closed to rational scrutiny. It is just such values that merit objective investigations into what they really are and do to reveal their real contributions to good.

A clearer concept of the relation of values to virtue can be seen by comparing the interests of society with those of an individual. Values, in the institutionalized sense, are the underpinnings, the foundations upon which rules are built for the purpose of defining the duties of an individual to his community. Values from the viewpoint of the individual may or may not coincide with those of his society. When they do not, then his personal values are the lights that guide him to think and act contrary to the wishes of his peers in accordance with the dictates of his conscience. The term value therefore implies an unstated social or personal authenticity. The absence of clarification lies as the nub of most difficulties encountered in discussions of values.

Virtue as moral strength returns us to values that sustain morality. Strength refers to virility that is derived, like virtue, from the Latin word *vir* meaning man. So it seems that virtue is not intrinsic either; it too is manufactured.

Values therefore are the utilities of knowledge; generally qualitative in aesthetics and morality, and quantitative in mathematics and the sciences, value as a utility in commerce being a mix of both quantity and quality of goods or services exchanged for quantity of money.

We shall discuss commercial worth (theory of value) under economy, moral worth as virtue, and artistic worth as a function of aesthetics.

Values as Perspective

We feel a compelling need to refer all values to a relational standard. A useful analogy is the location of bodies in space where all move relative to each other, but from any particular body, say the earth, all appear to be in motion except the base from which all else is viewed. All beliefs are similarly comparable to one's own that are presumed to be fixed. However, if we admit to having matured over time, is this not an admission that our values change over time?

Just as there is no absolute universal coordination system, there is no platform from which all values may be compared objectively. Values are local, to the point, subjective and purposeful. The values of contemporary society are fundamentally different from those of the Middle Ages, the difference arising from pursuit of truth as the core element of knowledge.

Distinguished from other judgements, 'value judgements' are not true or false per se, they are exigencies intended to reconcile contingencies with beliefs. They are true (meaning correct) to the extent that they accomplish the intended ends.

The true value of values resides in its foundation in truth. From truth flow the disparate values of freedom of thought and speech, dissent, justice, honour, human dignity and self-respect. Without truth freedom is free only for some, dissent is held to be indecent, justice is just ice, honour is ornery, dignity is ignominy and self-respect self-deception.

While truth is an essential constituent of value, we have come to understand it as an independent value, a subjective reality in holding values, distinctly separable from value itself. In the confused concept 'true values', true is only valid in the above context.

Personal values, and particularly the maintenance of respect for the values of others, make up our sense of human dignity and put civility into civilization. We should bear in mind that often it is not questions or answers that are important to speakers, but the opportunity to have an audience for a moment. Talking in this connection is a vehicle for the self-expression of a need: the need to be recognized. The contents may be secondary or of little consequence. The practice of sampling public opinion by polling serves a psychological function. It is not reasonable to expect the average person to respond thoughtfully upon demand to a substantive question. We therefore have to discount the answers to such sampling for what they say, and recognize them for what they are: opportunities for the expression of the id. They are 'id-entity ops'. From this point it is an easy hop to the conclusion that democratic decision-making founded upon such superficial levels of public expression is not likely to contribute to meaningful results. Such an understanding brings us back to the question as to who each of us is at any particular time and place as determined by our values. If we allow our values to be subverted by curbside questioning that solicits instant, glib, sound-bitten answers, then we become, at that moment, equals of our trivial answers. The presumption of entitlement to answers to all questions has become an impulsive characteristic of our culture, one that we fail to dissociate from the determination as to whether we should or should not question or answer. This presumption of entitlement is a product of the confusion of an interest in the pursuit of truth on the one hand, with the propriety of being, and allowing others to be, human and true reflections of their deeper values, rather than shadows of their shallow ones, on the other.

Values in Perspective

Finally, to close this discussion, we are obliged to conclude that while values are the personal property of individuals, principles are the governing directors of the universe. To question which is more sacred is to raise into question the propriety of one's personal values.

121 Excellence: Imminent Perfection

In entering the search for truth we are acknowledging, first, that we have not secured it and, second, that we are committing ourselves to a course of action that will hopefully draw us closer to attaining that end. The chain of thoughts that leads us in the pursuit of truth starts from the premise that we are governed by values and that such values are positive.

Goodness

Goodness and self are generally held to be complementary. Goodness fits the image that we have of ourselves being good: my goodness, goodness me!

We venerate goodness in others often after they are dead, but in the living in the manner in which we address others as, for example, 'my good man', often used in a patronizing manner. We exclaim 'good gracious', 'goodness knows', 'thank goodness' or 'for goodness sake', each, as if by implication, associated with a higher authority: 'God gracious', 'God knows' etc. So we personify or deify goodness to give it greater authority.

Goodness, as we have noted, has to do with appropriate or fitting relations in a particular situation. While goodness is acknowledged to exist to a degree in other persons and things, degrees of goodness are always subject to our own personal determination. Thus, goodness of fit means goodness of fit for purposes that benefit or befit one's self. The second test is whether that sense of 'fit' is appropriate for the purpose of confirming benefit to others. In this test of 'truth' we recognize a single operative principle by which all human values may be found to coalesce, our own goodness being reinforced in the views of others through our behaviour. The pursuit of excellence must be undertaken independent of economic need and greed.

Now let us apply these tests in a relative way to our aspirations to make improvements.

Excellence

In seeking excellence we commit to the search for improvement. The creed of the sceptic is belief in better ways and means. One has necessarily to ask the question: what constitutes betterment and how may we strive to attain such betterment? In correctly answering the question 'what?', we define the way; and in fittingly answering the question 'how?' we define the means.

The pursuit of excellence is improvement of goodness by degree, as a step or steps up a ladder of relative goodness from 'good' to 'better' to 'best'. The idea of excellence elevates the standard of one's goals. Remembering that difficult dictum 'the higher the fewer' concerning higher purposes of higher education, we return for a moment to the earlier statement to the effect that we venerate goodness in others in the manner in which we address them. So we do to those who personify excellence, addressing a Roman Catholic bishop, an ambassador, or a governor as 'Your Excellency'. The higher the stature the fewer the number of candidates we feel able to bestow such a courtesy upon.

Perfection

Our modern-day concept of perfection may be compared with Plato's claim that reality comprises ideal prototypes. Both constitute absolute standards as models which, like truth, cannot be attained in an imperfect world; both challenge us to disprove this thesis; and as experienced phenomena, both generate less-than-perfect semblances of perfection that we classify as poor, good or excellent. The distinction that we make is that we hold our experienced simulations to be real and the models to be ephemeral, whereas Plato held to the opposite ideals.

Whether we apply our talents to the production of models of 'myopia' or, as Thomas More did, of 'Utopia' (a word he coined in describing an ideal state of that name), what is essential to any such endeavour is the indulgence in innocent beginnings, and the necessity to constantly refine and redefine distinctions between the prototypical and production-typical models.

122 Health: The Maintenance of Vital Functions

We return to the topic of inertia to understand the starting point of our interest in health. We have no choice in the incidence of our birth. Birth is a consequence of chance and choice relations between our parents. We enter life through blood, sweat and tears. Having arrived, however, we subconsciously identify with life as all that we have. The emergence of consciousness confirms this instinctive understanding. Inertia dictates that life 'wants' to extend in perpetuity.

Whether the condition of our lives is tenuous, poor, good or excellent, we yearn for continuity of living more than any other desire. At beginnings we are helplessly optimistic. We believe in the impossible. We have faith that good will prevail and that we are somehow personally exempt from the misfortunes that overwhelm others.

Immortality is a manifestation of perfection. Slowly we learn to understand that immortality is an idyllic state that, like truth, is denied us in the absolute. We learn to accept that excellence is as close as we can get to perfection. Inevitably the forces of entropy will overcome the forces of inertia. The best that we can hope and strive for is an exceptional extension of our lives under the most favourable conditions. That is what our life's struggles seem to be all about.

We are not good for much if we are not good for much. Much depends upon our fitness to meet the challenges that our life's struggles present. To the extent that we are well equipped, we are well suited to prevail over adversity. Heredity provides us with a basic kit of parts. How well those parts perform individually and in relation to each other conditions our potentials.

Our bodily and mental performance is measurable on a comparative scale with physiological and psychological norms. Performance and integrity of parts acting in unison determine what we can do.

The Doctrine of Care

Remedy is the natural and desirable outlook of those concerned with the loss of vitality attributable to sickness or injury. Those that make such concerns their life's commitment operate in the discipline of medicine, the science of life maintenance.

At its inception the practice of tending to the special needs of the sick and injured was by means of trial and error. In the absence of knowledge of how the body functioned, curative means were unknown. The event of a cure was regarded as miraculous. The services of those who had previously been instrumental in bringing about such miracles were eagerly sought. Because the results of their services upon occasion were beneficial while their processes were obscure and thus appeared to be magical, the title witch doctor was bestowed upon them. In the light of today's knowledge we might more appropriately call them wish-doctors.

A prerequisite for working in any field is knowledge of precedence, i.e. what has been learned from past experience, and what conclusions can be drawn from such experience.

The State of Homeostasis

The essential quality of health is balance, known to the medical world as homeostasis. How well one is balanced determines how long one is suited to survive, all other considerations governed by chance being equal. Homeostasis is the tendency towards maintaining stability of living functions in the face of variations imposed by the environment.

When we review the performance of other life forms we discover that some have life spans just sufficient to ensure the passage of their hereditary messages to the next generation, while others manage to extend their lives through many reproductive cycles.

When we examine plants, the equilibrium maintained by, for example, a sunflower, a so-called 'annual', compares most unfavourably in life expectancy with a giant sequoia tree, specimens of which have been estimated to live for about 4,000 years (as determined by a count of their annual growth rings), making them contemporary with the builders of Stonehenge. This comparison is interesting because they both exhibit extraordinary growth rates that suggest that they are well matched to their environs. What determines that a sunflower may grow to a height of fifteen feet in one year and expire, while a sequoia in its early years may grow at a rate of up to eight feet per year to eventually gain a height in excess of 300 feet?

In the world of creepy crawlies, what enables aquatic rotifers to survive for perhaps one or two days, a fruit fly between one and two months, a bat between one and two years, while a giant tortoise may live between 100 and 200 years? The limiting factors may differ by species but in each case life expectation will be governed either by genetic or environmental variables.

The Physiology of Normality

A young man in the Bulgarian army was awaiting a medical examination, having been drafted into service. Ahead of him in the line was a man who was receiving extraordinary attention. There were no obvious anomalies that attended the condition of this man and to the trained physician and layman alike he appeared to be quite normal. If it were not for an astute examiner this man would have warranted no further consideration.

The 'problem' for this hapless man was that the standard examinations revealed that he had in fact no anomalies. He was the first statistically 'normal' man that they had ever seen!

Physiology is the discipline whereby models of physical normality in living organisms are established, against which all specimens of a like kind are compared for deviations.

Self Shelf Life!

Through the purgation of alien matter, the restoration of tissue and on to the substitution of organic and synthetic parts through the auspices of maintenance, repair and replacement, the extension of life expectations is advanced.

The Prolongation of Life

With the accumulation of a large body of medical knowledge curative means become diverse. Their effects can be compared. Over the long term some deliver more reliable, durable results than others. The notion of endurance is endearing. The development of strategies for extending the span of healthy lives is surely an admirable goal!

Man, with his propensities for mentality and meddling, has extended his life expectancy from about twenty years for Neanderthal man to over 70 years by the close of the twentieth century, largely due to his determination to create the most favourable conditions necessary to do so.

Natives of the Mediterranean island of Sardinia attribute their exceptional longevity to obsessive moderation, and, more recently, to good genes! Can we anticipate the discovery of a 'moderation gene' that would be amenable to general acculturation? Do we sense shades of a higher purpose?

Organic Health

The individual's interest in health stems from his broader interest in pursuing his other interests. What we refer to when we speak of health is organic integrity, the critical vitality of each part of a body upon which the beneficial performance of the whole depends. Impaired health is a state of reduced capacity to contribute to a system of mutually dependent parts. Critical health is the state of pending loss of performance in an organ, the function of which is critical to the continued viability of the whole.

While an organism by these definitions does not necessarily require to be 'alive', all life is organic and all non-living organisms mimic life in the sense of being dependent upon their separate parts or organs, where 'living' serves as a metaphor to describe what the organism does.

Social Health

Essential to the organic integrity of the one is the integrity of the many, for we depend upon plurality for reproduction. Denied the perfection of immortality, we emulate an idyllic state in copulation, understanding or not that while we cannot claim the secret of unending life, we can pass on our messages of life to those whom we hold most dear, for them to do the same. In the interest of affecting these passages and safeguarding the lives of those to whom we entrust our essences, we transfer our interests in self-health to the health of others. In this way each organism is eclipsed by a greater organism that is the sum of multiple lives forming categorical species, above and beyond which lies an ever greater organicism: the sum of all living species.

Systemic Health

We are led from each idea of organic integrity to another, larger in scale, upon which the lesser depends, and so on up the scale of scales. The health of the greater environment that enables the support of species and individuals is dependent upon the lesser in order that the greater may depend upon it.

Organic integrity depends for its claims to quasi-immortality upon the integrity of all that is smaller and all that is larger. The earth is a vital part of the solar system, that system a member of the Milky Way, in turn a small part of what we call the universe, which, because of our shortsightedness we presume to be the full extent of existence. What keeps the universe in equilibrium or good health if it is not in turn part of an as yet undetermined greater reality?

Through our chronic disposition to accrue greater knowledge we have painstakingly convinced ourselves that ignorance is not bliss; yet, to the extent that we do not know what we do not know, relative ignorance endures as relative bliss! Thus, we tend not to concern ourselves about poor health until we have evidence to that effect, or about mental health until after we have lost it!

Revitalization

We recognize the cooperative potential of chemistry to pair harmful and beneficial microbes, often in symbiotic relation, such that the detrimental effects of infectious diseases can be neutralized to afford protection to an organism from further damage or to extend immunity to future exposure. We acknowledge the biological processes whereby damaged tissues can be repaired by natural processes or artificial aids.

The understated question is whether mutation, being an aberration, is not the effect of, or of itself, a manifest disease. If so, a secondary question: whether disease, being normal, may not be wholly desirable as a means, perhaps the only means, by which we may make progress in our abilities to adjust to an ever-changing environment. If so, are not such diseases that work in support of modifying life to enable it to better withstand changes in the environment surely harbingers of evolutionary progress, forces for change without which we cannot survive as a species? Or do we reserve the term 'disease' just for those conditions that we perceive to be to our short-term detriment?

A further question then arises as to whether consciousness and intelligence are two such 'diseases'. Are they not aberrations in the context of 'normal' animal behaviour?

And beyond: are not consciousness and intelligence organic mechanisms for the development of still further means by which to self-repair, modify and improve the performance of our species in its search for perfection and immortality?

Bearing in Mind

Medicine, the most humanitarian of all professional callings, has focused since its inception upon matters physical. The challenge of tomorrow is to address with equal commitment the realm of the human psyche in search of homeostasis in mental function. The field of psychology concerns itself with the study of mental normality in living organisms, specimens of which are compared for deviations.

Psychiatry, the discipline concerned with treating mental, emotional and behavioural disorders, is today just emerging from its equivalent state as wish-doctors. Trial and error is still standard practice. The search is on for a means by which to dispense equilibrium as the panacea prescription for peace of mind!

123 Improvement: Progress Towards Betterment

What is the role played by the tendency for man to seek improvement, and why should he wish to do so? The answer has already been intimated.

The Tendencies of Evolution

Evolution has recognized the stream of events that we register as changes, and favoured those organisms that are better equipped to contend with their changing environments by according them higher probabilities of survival. Man desperately wants such changes as shall occur to permit him to continue to be included within the class of survivors. An infinite amount of conservative wishing for a state of relations that is constant and unchanging, as represented by the ideal of inertia, is not going to change the greater force of entropy that dictates that change will continue to occur. Evolution seems to confirm the veracity of Wallace's original premise concerning the tendency of varieties to depart indefinitely from the original type. Indefinitely does not refer to whether or not, but to what, where, why, when and how departures will take place.

The Sceptical View

There is a certain aspect of constancy to the notion of departing indefinitely from which the conservative may draw comfort. With consciousness working in his favour, the sceptic takes the position that because of his consciousness he is more acutely aware of the changing nature of his environment than other life forms, as a result of which he is better suited to adjust to changes and therefore to survive. That betterness is predicated not merely upon his awareness as an observer but upon his evolutionary gift of choice. Choice makes finity out of indefinity.

If, as Wallace contended, change departs indefinitely, we must construe that it will continue to depart. The question then arises in what direction does change proceed? The presence of the faculty of choice implies that, anyway to some degree, we may alter the course of events that would otherwise depart indefinitely had we not acquired such faculty. In other words, we have at our disposal the ability to replace a little indefinity with a little finity. The sceptic asks what next, meaning what do we want to happen next, and prepares for that determination – for the better.

Now is not the end. The sceptic always tries to extend the present into a future that will be better.

Wait, one might say; half the sceptics that I know are negative. These sceptics always point out the flaws in other people's ideas! This is true, but defining what, where, why, when and how something does not represent an improvement is all in a day's work in whittling down a residue that contains what may, should or will advance improvements.

Prospects of Greater Security

We are now led directly back to our discussion of security. Better means more secure. The search for improvement seems to have become a chronic condition in mankind. It therefore behoves us to diligently search for and accept rational definitions of, and paths to, improvements as the bases upon which to proceed from the present into the future, and to implement them to the greatest degree feasible in the interests of the greatest numbers in the name of planetary security.

Every step towards progress is achieved by overstepping existing bounds. As we have discovered, improvement in performance is a measurable process of incremental advancement step by step.

124 Corruption: The Debasement of Integrity

Impersonal Corruption

Natural processes continue to recur interminably; their common product is 'change'. All change is corruption of relations the effects of which are changes in capacity for doing. Change is a loss of integrity of a particular state of being and the assumption of integrity of others. Since all such changes, losses and assumptions occur over time, and since all things are in a state of change, integrity may be viewed as a 'snapshot' of a particular condition of a particular thing in a particular place at a particular time. Governed by such a definition, even chaos may be held to have integrity, moment by moment, 'frame' by 'frame', that defines its essential nature. Such corruption as we have so far defined it is debasement of integrity consequent to unintended natural causes.

Manufactured Corruption

Corruption as a concept is man-made. What we call integrity is by definition a definition. Some situations exhibit integrity, some corruption, all as determined by perception. What we refer to as corruption as the term is generally used, is the effect of negation of natural or civilized order by default of man.

In 1962 Rachel Carson's seminal work *Silent Spring* sprung open a dimension of public consciousness to which we had hitherto been blind. In 250 pages of a cheap paperback (a good read on a five-hour flight from San Francisco to New York) Carson unleashed a deluge of stunning indictments against human carelessness, greed and irresponsibility, the combined effects of which threatened, indeed promised, the extinction of life on earth if the conditions were to continue unabated.

Before Carson the discipline of ecology was widely regarded by scientists as a non-science for dilettantes and was relatively unknown to laymen even in name. The social upheavals of the 1960s in the US (when *Silent Spring* was freely quoted with the authority of a bible) were to change this radically. Ralph Nader's heroic 1965 criticism of the car industry in *Unsafe At Any Speed*; the strident environmental advocacy of Paul Ehrlich in his 1968 bestseller *The Population Bomb* and Timothy Leary's shenanigans as the self-appointed dean of the 'turn on, tune in and drop out' drug counter-culture arising in indignation over the US involvement in Vietnam, all signalled significant social reorientation.

The ecological awakening, consumer protectionism, grass-roots peace advocacy and the civil rights movement were often viewed as competitive forces for media attention, political support and economic priorities; however, together they constituted the critical mass necessary to catalyze the most profound social changes of the twentieth-century, having as their common theme 'anti-corruption'.

In a few short years ecology and consumer protection were to assume primacy in domestic and foreign policy politics. Since corruption had a new face, logically anti-corruption should also. The rights of people and peoples were to be further defined and legitimated.

There followed the spawning of a plethora of legislative, administrative and judicial enactments: the 1964 Civil Rights Act, the Clean Waters Act of 1966, the founding of the Environmental Protection Agency (EPA) and of the Occupational Safety and Health Agency (OSHA) in 1970, and the landmark US Supreme Court decision in the 1973 Roe vs Wade case that protected a woman's rights to govern her own procreativity. A burgeoning number of international treaties aimed at redressing the more pervasive and obdurate ills of human populations was initiated.

Ironically those who we refer to as politically conservative became the most reluctant to foster the conservation of natural resources or to recognize the necessity for placing limits on growth, production, consumption and waste, all of which appeared to conflict with the traditional business of exploiting whatever was there to exploit.

Personal Corruption

The era of the 1960s is the benchmark from which we can make distinctions between the unwitting and wilful debasement of the integrity of the human biome. The 1960s let it be known to all men that despoilment of the environment would be interpreted as being intentional and would be met with harsh retribution.

We shall deal with the subject of corruption as it pertains to damaging the interests of others with premeditation and malice when we address the subject of 'vice'.

At the close of the twentieth-century man was so attuned to thinking and acting in his short-term interests, as conditioned reflexes, that those who did project the consequences of our short-sighted behaviour despaired of ever stemming the tide of detriment. And yet we should not be surprised, for mankind is of the animal world, all of which acts in this way. It is only man who plans in the longer term at all, and he is slow in gaining understanding. What worries those who worry about the future is how and whether mankind can restrain the momentous forces of exploitation, waste and despoilment and restore himself to the otherwise self-balancing systems of equilibrium within nature.

Humanity has to revert to a condition where it no longer threatens the continuance of other life systems and resources before the earth's integrity is compromised beyond the threshold of its viability as a life-supporting system. The means to our ends, in order of preference, will be voluntary restraint, restraint under duress, restraint by force, some combination of the above, or extinction by default.

Social Corruption

Lord Acton's dictum 'power corrupts, absolute power corrupts absolutely' is writ dark from cover to cover in our history books. In contemporary life the application of the unbridled power of money and authority, by government and the private sector alike, gives rise to abuse of even the most sacred national treasure: the constitutional rights of the people. Unless comparable amounts of money and energy are applied to overcome such abuses, corruption rules every day of every life of every citizen. Coercion and extortion are the means, the extension of power and privilege the ends.

In the early thirteenth century St Francis of Assisi took the moral high ground to institute a cult of poverty in opposition to the corruption that he observed wealth led to. His efforts were to no avail because by that time the most powerful corporation in the land, the church, was a major player in the international banking system. We still bind God to the currency as if to lend vicarious respectability to the flip side of virtue.

The remedy lies in mediation between the interests of the rich and powerful and the interests of the impoverished. But how does a weak man mediate with a strong one? What leverage can he bring to bear? Numbers! Whether there is in place a system of government based upon democratic principles or of mob rule, numbers count! A second course of leverage is to institute absolute limits to absolutely everything, all proportionate to the potential that anything has for subverting the wellbeing of the people and their environment. We have trade unions, general elections, graduated income tax and a profusion of other checks and balances in our system of social organization. Why are they not working in support of the man in the street? Well, the sad answer is that they are; that is to say, they are enabling the rich to get richer, the middle class to grow thicker around the middle and the lower classes to lower their hopes and expectations.

The best indicator of how well a society is functioning is not to feel the economic pulse of the rich and famous, but to gauge how the citizens relate to each other. When the life experiences of different segments of the community have diverged beyond recognition that they are parts of the same social organism, then there is tension. Today tension is in the air, everywhere.

The callous disregard for the interests of others is the fuel of corruption. It is time for atonement whereby the rich and powerful voluntarily commit to trading their self-indulgence for social indulgence, measure for measure - or for the weight of numbers to be brought to bear.

Environmental Corruption

If we attempt to classify those conditions that tend to detract from the maintenance of sustainable organic populations we are confronted by the rude awakening that we, the human species, are in large measure responsible for the depletion of other species.

Population Increase:

The primary cause that drives our miscreant behaviour is the unbridled rate of increase in human populations. A simple test of this hypothesis is to compare the effects of human activity upon the environment when the world population was half the 6 billion of the year 2000 (i.e. 3 billion in 1960) with the likely effects of a population double that of the year 2000 (12 billion predicted by the year 2050) assuming that we will continue to conduct ourselves much as we did in the year 2000.

We refer the reader back to our discussion of relations to be reminded that the most critical aspect of additions to human population is not the number of additional entities per se, but rather the number of additional relationships that are inevitably simultaneously generated with the addition of each new entity.

The number of single point-to-point relations between individual entities is established on the basis of the formula $n(n-1)$ where n equals the number of entities under consideration. Thus we can simply compute comparative numbers of interpersonal relations for populations in the years 1960, 2000 and 2050 as follows: 1960: 3 billion people generate 9 billion billion relations; 2000: 6 billion people generate 36 billion billion relations; and in 2050: 12 billion people will generate 144 billion billion relations; where one (US) billion is understood to be 1,000,000,000.

This does not take into account that inter-personal relations are two-way and operate within multiple domains that include, but are not limited to, the temporal, spatial, social, physical, economic and political environments of which the most critical may be found to be the psychological.

Nor have we taken into account the nearly infinite number of relations established between each additional individual and all other existing impersonal things, most conspicuously evident in the elevated levels of competition for resources, not to mention aforesaid relations between relations!

The secondary causes of environmental corruption may be ascribed to just three human predilections: the depletion of natural resources, unrestrained land use and pollution.

Resource Depletion:

Resource depletion refers to the reduction in reserves of living and non-living stocks at rates of incidence that exceed replenishment rates. On land the conspicuous example is deforestation, including consequential losses attributable to islandization, whereby small 'islands' of the natural environment are preserved but isolated from each other, thus denying free-range migration and settlement opportunities for animal and plant progeny. The ultimate depravity is notoriously illustrated by the denudation of Easter Island. In the oceans the over-harvesting of certain species of fish render their prospects for recovery in the long-term tenuous at best.

In the world of the non-living the mining of rare minerals to exhaustion denies their application to unknown and possibly life-critical future needs; while the full costs of maintaining and substituting for existing petroleum reserves, if taken fully into account and passed on to the consumer, would translate to gasoline market prices in excess of 1,000 dollars per gallon!

Land Use:

Like mineral resources, land is a finite asset. Burgeoning human populations are generating unprecedented pressures to convert productive arable land to non-agricultural uses when the fact of their compounding numbers demands greater production of food and other replenishable crops. These losses are exacerbated by the dedication of large land areas to growing crops that not only offer zero sum benefits to society save the accrual of short-term monetary gains for narrow vested interests, but detract from the health, safety and welfare of the people to whom they are disingenuously targeted. Tobacco is an obvious example in the US, the 1978 production of which exceeded that of dry beans measured in metric tons. A comparative cost/benefit analysis of the real values of the two crops to society would embarrass all but the most callous capitalist!

In 1987 land uses in the United States other than for forests, meadows and pastures and agricultural lands under permanent cultivation constituted 19.8 per cent of the total land area. By 1997 this 'other uses' category had climbed to 30.9 per cent, a loss of over 400,000 square miles of productive land, or 11.1 per cent of the total land area of the United States, in a single decade!

Pollution:

The worldwide misappropriation of industrial products and by-products has drastically destabilized the natural environments of the atmosphere, hydrosphere and geosphere on an immeasurable scale. Whether by defective design, negligence or accident the release of chemical contaminates, waste particulates, noxious gases and radioactive particles has contributed to the premature death, deformity and sterilization of countless individuals of human, animal and plant species, members of that 'fourth estate', the biosphere.

Pollution can be characterized as effects of the inadvertent loss of control of harmful industrial by-products. A diatribe of misconduct and a list of contaminant pollutants are entirely inadequate to convey either the pervasive nature or the perverse effects of this phenomenon. Complex chemical reactions interact within and between the realms of air, water and soil to obfuscate our attempts to neutralize their pernicious effects.

We will briefly identify a few of the principle players and the forms by which they make themselves evident, understanding that such a summary account represents little more than waving a small red flag to alert those concerned, as we all should be, about the enormity and severity of the problem.

Atmospheric Pollution:

The atmosphere is perhaps the most conspicuous medium through which we become acquainted with mankind's malevolent relation to his environment. Atmospheric pollutants are diffuse. They may be categorized by generic type, chemical affinities, means of disbursement, their effects upon the biosystem and in other useful ways. We will confine this discussion to a short general narrative addressing causes and effects.

The combustion of fossil fuels and deforestation have queered the natural balance between supply and demand for oxygen, pumping up the quotient of carbon dioxide in the atmosphere (approximately five times the amount exhaled by humans) while diminishing that of oxygen. The symbiotic relation between plants and animals has been disrupted. Add to these emissions carbon monoxide, nitrogen oxides and sulphur dioxide and you have identified the final bequest and lasting effects of the twentieth century's most revered and least coherent designs: the internal combustion engine and all its oxygen-gobbling cousins.

Exotic chemical wastes are disgorged from manufacturing processes that carry the nefarious distinctions of accumulating slowly over time (like lead), moving through the environment like a chameleon (cadmium), or making direct assaults upon the very tissues of life (like arsenic). It is often only after considerable impacts have been sustained that we are able to recognize that what was hitherto thought to be essentially innocuous was in fact noxious.

Add to these those products that we indiscriminately release into the air, like DDT (dichloro-diphenyl-trichloro-ethane), that accumulate in animal tissues, pass from organism to organism and generation to generation and poison and kill in critical concentrations. Now release a generous quotient of CFCs (chlorofluorocarbons) used as propellants of pressurized liquids. These contaminants are not absorbed by biological tissues, remain stable and rise to the stratosphere, where they destabilize the beneficial action of the natural greenhouse gasses.

Finally (an ominous double entendre), we must admit to the contributions, real and potential, accidental or malevolent, of airborne radioactive contaminants misappropriated from nuclear technologies, the dangers of which the catastrophic malfunction at Chernobyl in 1986 should be our constant reminder.

Water Pollution:

While great efforts have been expended by the developed nations to nullify the effects of water pollution from identifiable industrial, institutional and residential sources; the control of contaminated run-off into surface water systems from hard surfaces, agricultural lands and aerial deposition has proven to be a more daunting challenge.

The over-drafting of fresh water from streams, rivers and lakes for agricultural, manufacturing and consumer uses has signalled limits in capacity that are slowly coming into focus as the subject of foreboding political controversy in upcoming decades. Critical shortages of potable water exacerbate the detrimental effects of pollution; the former impairing new growth while the latter vitiates existing life forms.

The incidence of marine disasters involving the release of crude oil, the dumping of industrial wastes, the discharge of raw sewage and the disposal of radioactive wastes in the oceans of the world continues largely unabated.

The bounds of the hydrosphere are indefinite insofar as water is held in suspension in the atmosphere and is absorbed deep into the earth, where its presence is ultimately subsumed by its absence. But in general, and in the absence of a clear point of origin, it should suffice to say that water precipitates from the atmosphere to fall either directly into larger bodies of water or upon land where it percolates down into aquifers, or sheets and streams into rivers from whence it flows into lakes and oceans, ultimately to be lifted back into the atmosphere through the evaporative process, to recycle ad infinitum.

We are reminded that the ocean 'is the pool of all'. That 'all' by any definition includes all the inimical pollutants that in their course contact and contaminate members of all species of all plants and all animals.

Soil Degradation:

Concomitant with land usage runs the issue of land abuse. The practice of overburdening agricultural land with excessive quantities of fertilizers the long-term effects of which sterilize otherwise fertile soils has led to retirement of arable land from productive use.

In critical areas erosion of fertile soils has assumed such proportions that annual losses far exceed the weight of harvested crops and the ability of the land to regenerate fertility; action that eventually leads to economic exhaustion. The loss of fertile soil translates to siltation in the waterways causing changes in the patterns of flow, reductions in dam capacities and to changing surface levels that generate new economic demands for maintenance.

The needs for disposal of solid wastes generate further demands for the dedication of lands for unproductive purposes and raise the prospects of contamination through the leaching of chemical contaminants into underground water systems.

Even more disturbing is the practice of stockpiling radioactive nuclear waste materials.

Dire Consequences

Smog and Acid Rain:

The unrestricted burning of fossil fuels coincident with natural inversions of atmospheric temperature (where temperature increases with height, usually at night) was attributed to have caused over 4,000 deaths in London from respiratory disorders during December 1952.

Atmospheric nitrates and sulphur dioxide emissions, primarily released from power stations, have accelerated the decay of building façades, the collapse of forest structures and the poisoning of lakes, causing inestimable economic losses.

Toxic Waste:

It is inconceivable that there can be zero-risk associated with the storage of nuclear waste products. The corollary, that contamination may be caused by negligence, depravity or natural cataclysmic events bespeaks of the unspeakable: the uncontrollable irradiation of unknowable populations and biomes, justification enough to phase out this industrial process for all time.

Global Warming:

In our earlier discussion of the atmosphere we touched upon the subjects of 'greenhouse gasses' and the ozone layer that afford a protective shield to the earth that is critical to the viability of life. We are now confronted by gradual but accelerating rates of over-production of greenhouse gasses, notably carbon dioxide, the effect of which is to render the earth's blanket a more effective insulator and reflector of long-wave infrared rays that would otherwise escape into outer space. In consequence the earth's surface temperature increases causing accelerated rates of evaporation and additional quantities of water vapour, a greenhouse gas that augments the work of carbon dioxide by falling back to earth as rain to release its captive loads of thermal energy.

The consequent phenomenon of climate change, while geologically natural and variable over tens of thousands of years, has been exacerbated by atmospheric pollution, the effects of which have been to increase the average global air temperature by 0.6°C (1.1°F) during the twentieth century. Projections for the twenty-first century suggest that by the year 2100 air temperatures might rise a further 1.4–5.8°C (2.5–10.4°F).

A direct effect of rising air temperatures is the progressive melting of glacial ice caps that contribute to a net rise in sea levels and a greater frequency in the incidence of cyclonic storms. The implications of a general rise in global air temperatures and sea levels are ominously portrayed in a flag. A lone sea bird flies in a red sky over troubled waters searching in vain for Kiribati, a group of low-lying coral islands in the Pacific. Today the population of Kiribati is about 90,000; but the significance of its precarious future is as the tip of a chilling iceberg.

A correlation between sea levels and air temperature is estimated to approximate a 15cm (6 inch) rise in mean sea levels per 1.0°C (1.8°F) increase in average air temperature. Taking the median projection of a 3.6°C (6.5°F) rise in air temperatures by the year 2100, the mean sea levels would rise an estimated 54cm (21.6 inches). If the high projections for increases in temperature were to transpire, the surface of the world's oceans could rise more than 1 metre (or 3.5 feet) by the year 2100.

The effects of rising oceans upon human demography compound geometrically. While the world population living within 3 metres (10 feet) of the mean high-tide level is uncounted, a single sample, those living in the Nile delta in Egypt, constitute about 20 per cent of that nation's population, or 13.2 million people.

Worldwide costs of the relocation of populations to higher ground, the building of new infrastructures, and losses attached to the inundation and salination of agricultural and forested lands are incalculable. What can be interpolated is an economic impact that would incapacitate many existing national economies.

From a geopolitical viewpoint two quantitative questions arise: how many people within existing economic units (nations) would be adversely affected, and how much is that figure as a percentage of each total population? Projections into the prospects of a few conspicuous trouble spots are revealing. Under the worst scenario, 72 million human lives would be adversely affected in China and 71 million in Bangladesh. At the time of the projections those numbers translated to approximately 6 per cent and 60 per cent of their respective total populations.

By contrast the 280,000 people of the Maldives, 90,000 in Kiribati, 83,000 in the Seychelles, 10,900 in Tuvalu and those in other low-lying island groups, would be adversely affected 100 per cent.

A more pervasive and potentially greater peril of global warming is the prospect of the failure of animal and plant species to adjust to traumatic lifestyle changes or to migrate to and propagate at higher elevations in climatic zones newly made equivalent to those to which they were accustomed but that were no longer hospitable.

What could be more ironic than for the island of Mauritius with a population of 1.2 million to sustain the same ill fate by dint of human carelessness, greed and irresponsibility, as did its best-remembered inhabitants, the now extinct dodo birds? Flags flag omens. The flags of all of the above-noted states fail to portray any vestiges of human or (excepting Kiribati) animal life!

Habitat Destruction:

The primary effect that flows from the three secondary causes of environmental corruption is a reduction in diversity and opportunities for reciprocity culminating in the destruction of habitats, with attendant losses of life.

Species Depletion:

The secondary effect of the three causes is the depletion of species, the ultimate conclusion of which, from our point of view, would be the self-induced extinction of the genus Homo Dodo.

Conclusion

The conclusion that must inevitably be drawn is that human beings, exhibiting consciousness, will and volition, are the only life forces that can, to some degree, comprehend the total reciprocal system of life on earth and are thereby the only prospects able to act protectively in the interest of its security.

Since we have no way of anticipating which individuals will rise above their peers to contribute more significantly than others, we must afford all people equal protection under our laws, not to protect them solely as individuals, which we hope to do as a means of protecting ourselves and those we love, but to protect the measure of collective potential that we have for supportive services. This is the most fundamental virtue of democracy.

If one considers the total impact of one's visit to earth we discover that we make enormous personal demands upon the earth and its resources. Consider how many cows and chickens each of us causes to be slaughtered in our lifetimes; how much fossil fuel we burn and what that translates to in tons of greenhouse gasses released into the atmosphere; how many trees have been felled in response to our demands for forest products. Granted, we have spent one-third of our time asleep, which, in the context of this discussion is tantamount to a 'dream'. The best possible thing that one can do to maintain the equilibrium of the planet is 'nothing'. It doesn't need us, it doesn't want us and it doesn't care. Any suggestion of the existence of a supernatural 'caretaker' of the earth does not pass any test of any rational demonstration to that effect. In the final analysis all is corruptible. Is the end of this story to be permitted to be the end of all stories? Is our collective epitaph to read that responsibility for the *terra cariosa* and *damnata* (rotten and condemned earth) shall be borne for all time by the unconscionable souls of dead men?

125 Restoration: The Reinstatement of Natural Balance

The unequivocal prime indicators of the health of the planet are the diverse life forms that it supports and the security of the ecosystems upon which they depend. For mankind the environmental detractors from these fundamental truths are largely of man's own devising and principally concerned with local economic wellbeing and security, the conflicting interests for which we pay dearly. Restoration is the process by which man seeks to recover from the effects of his poor judgement and detrimental actions.

Economic Aspects

The fact of competition between short-term economic gains and long-term environmental interests is man's locus of vulnerability. The conflict between conservation and the exploitation of natural resources illustrates this point well. If, however, man were to concentrate upon his long-term interests through sound strategic planning, satisfaction of his principle short-term needs would flow naturally. The reason that we do not may be ascribed to our egocentric character, understanding as we do that we are merely mortal short-term visitors with large appetites that hold positions with authority for making decisions in the short term.

As a result we attempt to mitigate unfavourable conditions by means very similar to those by which we create them, by placing responsibility for misadventures upon the persons perceived to be the most conspicuous or last in the line of causation, a remedy that does not address root causes.

For example, the manner in which we advance false strategies for penalizing industrial polluters fails to recognize and prevent those costs from being passed on to consumers. Where critical services and products are involved, the effect of such an ill-conceived remedial system is to force consumer costs up and thereby increase the incremental profit mark-up for the polluter.

Reliance upon Science

Science examines what 'is', in order to enquire what is going on. Only by submission to logical process can responsibility for corruption be rationally portrayed. Simple anomalies may be revealed through finding a one-to-one relationship between cause and effect. The relations of asbestos to asbestosis and of acid to acidosis are examples made more obvious after the fact of discovery by reflecting the name of the cause in the name of the effect. Such connections are, however, seldom simple and, where all contributory elements to a complex set of relations are unknown, the processes of testing hypotheses become seriously impaired.

With or without full knowledge, logic demands that risks that arise from awareness of multiple disagreeable experiences be analyzed on the basis of comparison. It is only from such data that priorities may be established with respect to determining which issues merit the allocation of finite resources.

Restorative Process

If we take today's environmental conditions as benchmarks from which we resolve to make progress in the direction of restoring relative equilibrium in nature, it follows that only by reducing detriment to levels below the 100 percentile datum can we make advances. A decrease in unitary contributions to pollution at source points does not of itself reduce detriment if the total number of source points is allowed to increase. Thus, absolute standards must govern total contributions to pollution reduction while unitary reductions become the means to such ends; absolute goals being predicated upon pre-industrial levels of approximate balance in nature. By such a strategy maximum allowable total emissions may be reduced proportionate to progress made.

The ultimate goal of energy conservation would appear to be to capture the excess energy attributable to global warming and to convert it to usable forms. To this end the most propitious socio-economic mechanism would be to retire the nuclear and fossil fuel-based power-producing industries concurrent with and proportionate to advances made in solar energy conversion. It is noteworthy that the earth receives more energy from the sun in one hour than mankind generates in one year.

Subsidizing innocuous renewable energy industries at the expense of industries that rely upon non-renewable and toxic resources could deliver risk-free clean energy to consumers at affordable rates within a generation.

Political Initiatives

The question of redressing local corruption can be managed within local jurisdictions presupposing that the will and means to do so are present. More often the perception of corruption is only evident when the means to a cure exceeds the capacity of local authorities to redress the disjunction. The burden of effort to make good is thus distributed not simply to contributors proportionate to their responsibility, for say pollution, but is extended to those beyond to the extent that their interests and participation are impacted and necessary. Whereas water and air pollution can readily be identified as being of global interest and thus merit the application of a coordinated response, the depletion of rare mineral resources less obviously affects others. It is only when total assets for a particular commodity or resource are appraised worldwide and then compared in relation to rates of consumption and replenishment that a sense of appropriate balance can be ascertained. Each act of corruption requires to be accorded a scale of magnitude from which an appropriate strategy for control and remission can be devised vis-à-vis the need for a local, regional, national, international or global response.

While no government commits to protecting the environment voluntarily, popular demand is generally sufficient incentive for any representative government to make that commitment. Ultimately, responsiveness to all such extraneous matters is gained through political means.

One such promising political initiative has been the 1997 convention of a conference on climate change in Japan under the auspices of the United Nations. Attended by 160 nations, an agreement was forged known as the Kyoto Protocol (on global warming). The agreement required ratification by the major industrial nations precedent to becoming international law. By 2001, 178 nations were signatories to the treaty, including the European Union (EU), Russia and Japan (second, third and fourth largest contributors to the emission of carbon dioxide). China, a less-developed country not required to ratify the treaty, had voluntarily done so.

The position of the US (responsible for generating 24 per cent of the world's carbon dioxide pollution) was initially equivocal, turned irascible and subsequently deplorable.

A controversial system of trading had been proffered by the World Bank whereby clean air technologies would be sold to less developed countries in return for the designated credits that each would earn for their contributions to reducing the production of greenhouse gasses that fuel global warming. The payoff to the industrialized nations: business as usual! Manufacturing jobs would be secured and dependencies created for the longer-term needs for maintenance, repair and replacement of the exported systems. The burdens of their obligations to reduce domestic pollution would be reduced proportionate to the credits earned without necessity for addressing the root causes or reducing pollution at their sources. Through such devise the industrialized nations could pay off their legal obligations to cease and desist from pollution with conscience money. The question therefore arises, what needs rehabilitating most, corrupt human habitats or corrupt human habits?

126 Security: Mental Quietude

What do we have in our minds, or more properly what do we not have in our minds when we feel the need to act in our own interests?

Security of Persons

What each of us values most is the fact of our own life. Whatever is required to extend that life beyond the present, each of us is willing to do with few exceptions. As individuals we perceive survival as our essential interest. In order to achieve this end, and unrelated to any imminent threats to our lives, we recognize our needs for food, clothing, shelter and medical aid.

Security of Relations

When we attain bodily security we are able to look beyond to those for whom we have responsibilities and with whom we share affinities. Putting emotions aside, these relations are our mutual lines of defence against danger. The larger the community, the greater its ability to secure its interests, all other things being equal. Social security is not simply the name of a government programme; it is the principal motivational force underpinning human settlement.

Security of Properties

Properties are means to ends, the primary end being security. Whether material support is controlled on a personal basis or by society, the measure of its primary value is the degree to which it wards off adversity and mitigates basic needs. The value of all property beyond performance of that function is academic or relative, being applied in the pursuit of happiness. Freedom may thus be defined as freedom from need to the extent that happiness may then be pursued.

Security Against Uncertainty

Underlying all our actions is a driving need to know. Knowledge is secured by certainty. But we have already established that uncertainty reigns, so what options have we? We can alleviate the extent to which we are disadvantaged by increasing the odds in our favour in the game of chance. What we feel the need to do is distance ourselves from the severity occasioned by uncertainty by stockpiling reserves against insecurity to cushion us against the worst contingencies. In so doing we reduce the immediate probabilities of misfortune. This is what capitalism does for individuals and socialism does for communities. On balance everybody needs the same degree of security from life's misfortunes, so how can anyone assert that they are entitled to privilege? If people are conspicuously disadvantaged, it is not privilege that they seek, but parity.

Security of Futures

We have earlier discussed the bearing of intelligence upon the relation between future security and probability as represented by a bell curve, and the pursuit of security as the principle motivation behind education. Therefore, with respect to security, time is of the essence. In furthering our interests in security we push the threshold of insecurity further into the future.

Security of Insecurity

To play with the unknowable (or perhaps more succinctly the incommunicable) is to exercise ignorance. To organize others for such purpose is to exploit their ignorance in the prospect of infusing a false sense of security. These are the destinations towards which mysticism and chemical dependencies lead. An understanding of the role that security plays in life cannot be gained without acquaintance with its antithesis, insecurity. Security of itself is an insecure concept in the absence of context. Insecurity is here to stay.

Security of Security

With each step in securing our persons, relations and properties, our need for security diminishes, or so our common sense suggests. However, there is a perverse anachronism to this truism whereby the accrual of individual security represented by wealth and power generates additional incentives to secure each increment of gain. A chronic disposition towards pursuing security, the product of a linear mindset, may be represented as a psychological defect whereby disparities between the worlds of the secure and the less secure are made more pronounced, generating resentment, enmity and corresponding mental insecurity contrary to original intentions.

Security of Mind

Security of our future prospects of security is the pre-eminent interest that underlies all our earthly concerns of which peace of mind is the most precarious. The question as to why people believe in things that cannot be demonstrated to be so still stands. The best answer appears to be because the act of believing delivers a sense of security. As to whether such security is real or false is secondary. The fact that such beliefs enable people to function with greater confidence in their judgements and actions are rewards enough, but not guaranties.

We all need a measure of security, whether in the form of sustenance, shelter or provisions to meet future contingencies. The big question is whether we allow the pursuit of security to consume our lives and the lives of others, or simply to free us to pursue other interests.

The Slippery Path

The key to achieving individual and social security rests upon the means by which society determines 'how' its interests in these objectives are secured, maintained and directed in the light of changing circumstances; a question that reduces to 'who' may be entrusted with such responsibilities?

127 Danger: Contingent Loss

The antithesis of security is danger. In order to understand what danger 'means' we need to consider, as we considered the topic of security, what the 'object' is to which danger refers, i.e. is the object a person, property, a mind, or some other 'thing' represented in the mind?

Danger as Perception

Danger is most commonly understood to be impending negation. There is no danger 'of' happiness, only danger 'to' happiness. If danger is impending negation, then to what is its negative force directed? This question cannot be satisfactorily answered until a question of greater precedence is addressed: by whom is danger perceived, meaning in whose mind? This is the 'where' question. Either danger is recognized or it is not, but the absence of a perception of danger does not negate the presence of danger.

Danger as Prospective Detriment

Of what and to what are negative effects referenced? What is it that is endangered? Danger is object-related. There is no danger in an idea per se. Any danger that we may attribute to an idea is in actuality the danger resulting from transformation of an idea (a theoretical danger) to a practical consequence. This is the 'what' question.

Danger as Probability

The severity of a danger is a function of the worst possible consequence and the probability of occurrence. Since danger is invariably a function of proximity (or projected proximity) whereby the closer the danger comes to the object, the more likely it is to affect it, this is another of those Newtonian formulae akin to the force of gravity being the product of mass multiplied by the square of the distance. This is the 'why', 'when' and 'how' question.

Denial of Danger

Lack of perception does not deny probability of impending detriment. A bump on the head delivered from the back is as much a bump on the head as if delivered with foreknowledge on the forehead. As to whether one is denied knowledge of impending danger or denies one's self-knowledge, the effect is the same.

Somewhere Between Perception and Denial

This is where most of us are most of the time. Danger is all about us. If we were to be aware of all dangers to our persons at all times, then we would not get out of bed each morning. So, with respect to knowledge of danger, ignorance is bliss for most of us most of the time. Only those dangers that portend to carry the worst possible consequences and the greatest probabilities of occurrence are critical to our perceptual needs.

Danger Contingent Upon What?

Danger is one possible effect of a cause. The function of perception is to offer through awareness the opportunity to mitigate the severity of detrimental effects before the occurrence of an event. Danger is not contingent upon perception, mitigation is. Loss is contingent upon the worst possible consequence multiplied by probability and divided by mitigation.

The Danger Equation

There are three possible consequences of perceived danger: that as a result of probability it has no negative impact upon an object; that it does have a negative impact upon an object; and, lastly, that the impact is reduced or eluded as a result of awareness and subsequent acts of mitigation.

Danger as a Pet Aversion

It is the perception of danger that threatens our minds, and the detriment caused by negative impact that harms our persons, relations and properties.

128 Industry: Human Activity

In discussing energy we made distinctions between 'energy', 'force', 'work' and 'power', all of which contributed to change. Our obsessive preoccupation with security dictates that we make efforts to advance such interests.

> The step beyond the thought is great indeed
> A spur to action in the name of need
> Oh justice can you try a little harder
> To weight the balance favourably for ardour?

Such commitment transforms passivity into activity. Doing involves change, change induces direction, direction implies purpose and purpose demands justification.

We have related the steps by which we secure security, from which we can deduce the tasks that we need to perform to gain such ends.

Work as Necessity

Our daily bread, protection against adverse climatic conditions and defence against aggression are minimal survival requirements that, when secured, enable us to turn our minds beyond the moment to contemplate other 'goods'. These are afforded through payments in exchange of value for equal value. The principle of equality is fundamental to working relations, whether as money for services, as equal willingness to exchange or as approximation of equality between the employer and the employed. Thereby the total value of work and payment are offset with mutual dependence of one upon the other. These are the first principles of cooperative work

Work as Obligation

The fact of dependency, whether of family upon a breadwinner or of a larger group upon breadwinners as a class, places obligations to work upon all who are able-bodied. In the larger society this obligation is formalized and formulated such that the greater the means, the greater become the obligations. This is the basis upon which graduated taxation is levied.

Compensation for work offers opportunities to secure all of the above, and to spend, save and enjoy a brief respite from the daily grind, enabling acquaintance with other interests for refreshment, a few moments under the sun, contemplation and time to give thanks. All such compensation is external and tangible.

Work as Pleasure

But there are other rewards for what we might wish to call 'rightdoing' that we shall discuss later in greater detail under the general heading of the affections; rewards that are internal to each of us and that derive from love, benevolence, gratitude, the conscience, magnanimity and virtue in connection with the act of working for others.

The act of commitment to exploit natural affinities and aptitudes stimulates our identification with work worth doing. In so doing security of mind is further advanced, a state of peace and tranquility serving as additional compensation. For the vocationalist, appreciating life and working are so interdependent that without the one the other isn't complete.

Remembering that doing involves change, change induces direction, direction implies purpose, all must be for the general good if it is to secure a deep sense of fulfilment for the individual. That is the justification of human industry. Work is the first path to happiness.

Work as Significance

Pleasure in work derives from the significance that we attach to it. Hesiod is credited as having invented the concept of work as an ethical undertaking, in stark contrast to the contemporary heroic ideals that we find references to in Homer's epic poems.

As the product of our application of force over time, work serves to inflect the direction of our efforts, thereby giving emphasis to a clearer articulation of our values, or by default, exposing our lack of worthy values. In either event, work works to alleviate the burdens of less favourable circumstances and to elevate the standing of the more fortunate.

If there is any merit to perpetuating inane activity it must surely reside in the unintended production of outlets for the innocuous relief of excessive mental and physical energy.

Productivity

The principal components of production are labour, capital, land and its natural resources, and organization. In order to increase productivity an increase in the supply of one of these components is necessary. Each of these considerations is discussed in greater detail under the headings of commerce, technology, collaboration, socialism, capitalism, economics and egotism.

In order for beneficial growth in productivity to occur it must be justified by a corresponding growth in demand, in the absence of which excessive production is counter-productive and translates to reductions in returns on investment, and waste. Beyond natural limits, productivity gains depend upon scales of organization proportionate to output capacities. An increase in demand exerts pressure upon improvements in production capacity that can only be achieved through economies of scale. Hence conglomerate industrial and commercial enterprise.

Labour, being generally in good supply, is in a relatively weak bargaining position vis-à-vis the capitalist, so what does the labouring class do to improve its conditions? It aggregates labour into unions to enhance its bargaining power, ostensibly monopolizing labour by which means it can withhold labour from industry pending the negotiation of equitable terms of employment.

What kind of behaviour is this if not intended to compensate for perceived disadvantages gained by opposing interests? Only where parties to an enterprise work in harmonious relations can they both reap just returns on their investments.

129 Behaviour: Environmental Responsiveness

For the purposes of this topic we shall confine our discussion of behaviour to human conduct, having explored many other existing forms and relations operating under the auspices of what we broadly accept as nature.

Sub-Human Behaviour

All living organisms less complex than man operate under strict environmental constraints imposed by the availability of resources and the incidence of predation by other organisms. In this regard no species has the capacity to eradicate any other. As the prey species diminishes in number the predator species suffers a corresponding loss, giving the prey species the opportunity to restock its numbers due to a diminishing number of predators. A balance between maximum and minimum numbers for each species is thus maintained within a range determined by the populations of species above and below each on the food chain. The herbivores' range is governed similarly by alternation between abundance and sparsity of vegetation determined by climatic variation, a condition clearly evident on the Galapagos Islands that are subject to El Niño cycles. Thus, the likelihood of any species becoming extinct is slight, occurring only under catastrophic circumstances.

Natural Human Behaviour

Man performs much like other life forms except that he further protects himself against contingent scarcities by cultivating stocks of lower organisms in sufficient quantities to maintain his needs for food and other essentials. He places limits upon the habitat ranges of those species with potential for preying upon mankind. If this was the full extent to which he differed from other species we might say that the added increment of security that he achieved had a relatively benign effect upon his environment and that his behaviour was inherently non-threatening to the survival of other species. But this is not so.

Unnatural Human Behaviour?

The emergence of human consciousness and volition represent anomalies in nature whereby man appears to have acquired a capacity to wilfully change the course of events that would otherwise have flowed from the operation of physical principles and probability.

Such appearances, while unprecedented, cannot be construed to be unnatural insofar as human life has evolved naturally from a simpler to a more complex organism. The anomalies arise from a quantum change of scale in man's capacity to cause and therefore effect change.

Notwithstanding his abilities to wilfully interpose his will upon events, man still operates within the environmental regimes prescribed by nature.

Man has conditioned himself during his survivalist years to prey upon animals that he no longer depends upon for sustenance. He purports to justify killing as sport, commercial necessity or survival expediency. Finding that such activities are culturally condoned, he presumes them to be natural and acceptable. The fact that in many cases he places his quarry-species in jeopardy of extinction and in all cases inflicts cruel and unnecessary suffering and loss of life is a reflection upon mankind's social mores and callous disregard for life in general. As long as human society continues to tolerate wanton killing, indulgence in this perverted behaviour will continue. How can we account for behaviour that is inherently threatening to life in general except by concluding that mankind as a class is still very primitive in its attitudes towards its environment? We still do not understand the delicacy of our relationship to other life forms as that of interdependence rather than dominion.

Man is a probability freak. His evolutionary path has afforded him a revolutionary capacity to interfere with nature's code of probability by allowing him to impose narrowly proscriptive sanctions upon her otherwise systematic and balanced regimen.

Economics as a Principle of Conduct

Insofar as man is never in full possession of all the facts, he confuses what he knows with what he does not reasonably know, pseudo-justifying his judgements and actions in beliefs. From this state of impropriety man proceeds to undermine not only his own species, but all those upon which he relies directly or indirectly for survival. Through the application of reason he puts himself in the untenable position of necessarily needing to do things that he hitherto conceded to God, the most significant of which has been to safeguard the maintenance of life on earth, ostensibly based upon probability as determined by adaptability and chance. *Noblesse oblige,* the traditional obligation of the French aristocracy to behave honourably, generously and responsibly towards those less privileged, meets with disdain and contempt in societies that are indoctrinated to compete and prevail at all costs and under any pretext.

If we return to the sceptical view for a moment, the now-familiar question returns: how can human society organize in a manner that results in behaviour consistent with the interests of all life forms, thereby enabling them to continue to flourish and mutually support each other?

While we have not treated behaviour as a psychologist would, we have revealed just enough to suggest that human behaviour is intimately connected with a state of continuity while remaining open to change. Normality operates to accommodate both ranges of possibility. For the human race it is normal to operate in a milieu in which there is a preponderance of ignorance over knowledge.

If so, how can we contend that we should overreach the authority of nature without pretending that God knows all and so directs? Are these positions reconcilable? Is our fate to be understood as being as predetermined as a *fait accompli*? If so, why bother?

130 Politics: The Non-Violent Promotion of Social Policy

Politicians, acting in accord, facilitate and disallow futures. Insofar as they advance ideas and values that exist, they are all conservative. It is safe passages into the future that they attempt to secure. They rely upon experience to tell them that certain developments are more beneficial or threatening than others, upon which they found their convictions and towards which they direct their energies.

Contingent costs and benefits are circumstantial in nature, about which we can only conjecture. So, as we discovered in discussing 'security' more generally, we become less and less certain, and less and less able to secure our futures the further we look ahead in time. We should not, therefore, be too surprised that the attention of politicians is directed towards immediate or short-term gains. This is reinforced by the knowledge that political authority is dispensed in short-term increments fraught with future uncertainties.

Axioms of Politics

Nominally we would expect that politics would be conducted in the absence of violence (when it would be called something else), be concerned with the general welfare, be conducted in an atmosphere of cordial civility, embody precepts of personal responsibility and embrace a commitment to serve the community. Within that envelope there is still plenty of scope for divergence of interest, understanding that it is only convergence of interest that can develop the critical mass necessary to dictate social policy.

The Political Spectrum

If we view politics roundly, we discover that it is well represented analogically by a clock face. Despotism stands at high noon, the pinnacle of totalitarian authority. At its polar opposite, six o'clock, reclines liberty, the patron saint of tolerant liberal thought. The right half of the face is the conservative encampment, moderating its solidarity as it moves from three o'clock towards six, and hardening its posture the closer it moves towards twelve, at which point it becomes absolute, and absolutely dictatorial. On the left we have the social conscience opposing the egocentric view across the divide. Like the conservative, the socialist mollifies his stance as he approaches six o'clock, conforming more stridently to ideological doctrine approaching his hour of absolutism. The clock analogue displays political options as radiant possibilities. In consensus politics, whatever persuasion is dominant at a particular time will, like time, change. Anarchy does not appear upon the face of the clock because it is not a political form, but rather the absence of one.

The View at Three O'Clock

Conservatism, that natural disposition to maintain a steady state, demands attention to the relation of surroundings to the self. This typically precludes a keen concern for the relations of surroundings to others. Thus the general perception of others towards conservatives is that they are excessively self-serving at the expense of others. The conservative is passionate in his conviction that man should not be tortured by unnecessary or rapid change. Life is good and that goodness should not be diluted. More of the same is all for the good he would argue. He fears loss of privilege and loss of opportunity. By and large, the more privilege people have, the less they feel responsible to and for society. The conservative frequently arms himself in defence of his alleged God-given rights to be what and who he is. Do you remember 'to be'? That was the essence of existence, identity and integrity. The posture of the conservative in a capitalist society may be characterized as being fearful and defiant with respect to change. The conservative intellect is used to militantly defend established values and to attack ideas that threaten them, whereas the progressive intellect investigates ideas in the abstract in search of benefits. Neither is intrinsically good or evil. They serve us well in different ways under different circumstances. It is the strident entrenchment of conservative and progressive beliefs that obscures their virtues to their opposite numbers. From the viewpoint of the rest of the world, the preponderance of American political convictions lies on the right-hand side of the clock face. The appeal of conservatism is its simplicity. That does not make it correct, but merely simple to relate to and adopt as an ideological panacea.

Where claims are made in the name of conservatism, the attending obligation is to ask what it is that is worthy of conserving. Conservatives may be compassionate towards conservatives, but they certainly are not towards the disadvantaged. It is easy to be conservative in one's own interest, but that is not the purpose of government. Governments are to serve the needs and interests of the total populace.

Nine and Counting

The socialist position recognizes that people are born unequal but maintains that justice demands that they be afforded equality of opportunity. Everybody should count, regardless of his circumstantial birth. The socialist is the Robin Hood of politics. He views progress as movement in the direction of equation of circumstance. The interests of the many override the interests of the few. Evolution prevails in time over conservatism, not just because change assures that it will, which it does, but because interest demands that it shall. The posture of the socialist in a capitalist society may be characterized as hopeful with respect to radical change. He seeks opportunities for meaningful participation in society for all.

Pendular Politics

In regimes that practice democracy, the pendulum of power swings within the arc contained between the five and seven on the clock face most of the time.

The irony in power politics is that when the balance of power moves from one side to the other, all views with respect to who fears and who hopes, and who represents the many or the few are reversed. So the terms 'conservative' and 'progressive' are simply convenient labels, useful in describing one political position relative to another. True conservatism has to do with safeguarding the infrastructures that enable life, political dialogue and due process to continue, rather than reflecting the interests of one party or the other. Real progress is continuity of movement in the direction of broad social benefits.

There is an aspect of man's psyche whereby he is attracted to things that he doesn't have and complacent when accustomed to having them. This attraction coincides with a principle of economics whereby scarcity generates greater demand. This becomes manifest in politics as alternations in the supply of, and demand for, individual and social opportunities. It is this transposition, like an alternating electrical current, that drives the turbines of social change.

Every small child knows that the further the pendulum swings in one direction, the further it will return in the opposite direction. Somehow immoderate politicians seem to overlook this principle, or are too fearful, self-centred or focused upon the short term to take it into account.

In accordance with the law of averages, the median position of the pendulum is always pointing down. Inertia propels forces beyond this neutral state of rest. In democratic politics the balance of power is generally held by those with a moderate disposition. However, we should be quick to recognize a measure of identification with most causes resides in most of us.

The High Wire Act

The conservative tradition would maintain that government should respect the individual in his rich variety, and that much of human value is no concern of the state. The socialist view seeks homogeneity in society, holding that only government is capable of securing its agendas. Somewhere between these views has evolved a liberal compromise: the 'general will'.

The liberal attempts to balance the interests of the individual with the interests of society. Equilibrium, nature's secret force, confirms the correctness of liberal thought and action as that most in synchrony with nature's predisposition to tolerate deviation in equal measure to left and right of a median position in its quest for just reconciliation.

When the threat of international communism appeared to be in recession, the forces of right-wing demagoguery, dependent upon the promotion of fear to gain support in adversarial politics, attempted to corrupt the term 'liberal' to connote socialist. The rebuttal is as close as the nearest dictionary. One such definition of 'liberal' runs: 'Broad-minded, open-minded. Not bound by authoritarianism, orthodoxy, or traditional or established forms of action, attitude or opinion.' Such a definition should bring reassurance that liberalism is the fulcrum of democracy.

Party Time

One's political disposition is determined in large measure by one's status relative to others. Reminded that mechanical advantage is gained by distancing a force from the pivotal point of leverage, and that additional force translates to additional power, individuals are inclined to subordinate their minor differences and to combine their energies in opposition to those whose views oppose their own. No amount of force can contribute to leverage if it is applied directly to the fulcrum point along the neutral axis. In societies that operate on the basis of democracy, political aggregations, known as 'parties', need to represent that their ideals more than offset the advantages offered by their opponents. The further their posture is removed from that of their adversaries, the more difficult they are to overcome. However, as party manifestos get further removed from the balanced views of the average, reasonable person, they attract less support.

What constitutes the good, who the intended beneficiaries are, how the costs and benefits equate with priorities and whether short- or long-term interests are being served – are the main grounds for contention between the parties.

Convictions represented by party politics function like accelerators and brakes with respect to governing the pace of change. So what we do in alternating rhythm is to speed up and slow down. The net effect over time is to proceed faster than some and slower than others would wish.

Three-party systems tend to be structurally more stable, to draw a geometrical analogy, since any two parties may align to triangulate and contain extreme behaviour exhibited by the third.

Independence

A truly optimal view of politics would necessarily be independent of doctrine and party affiliation. If one is to opt, for example, for excellence, then one must first be free to define excellence in each context under consideration, and then to act in support of each without conflicts of interest. One can do this as an individual, but not as a team player. Independence is a virtue rather than a strength. Compromise undermines virtue. One should not, as a matter of principle, exacerbate conflicts if one is elected, paid and committed to resolving them.

Political Neutrality

Political activity is not naturally congruent with anything else. It is often mean and insensitive. Today political polemics speak more eloquently to the deceptive use of words than George Orwell's 'doublespeak' ever did. Politics are 'polytricks'!

However, there are certain human interests and undertakings that are intrinsically non-political, the essences of which are incompatible with political process. In the world of the arts and letters, where freedom of expression is fundamental to creativity, any external imposition upon such work undermines its integrity. Similarly, fundamental science searches for truth in knowledge. There is therefore a legitimate need to protect the independent integrity of innocuous activities from arbitrary political interference. The governing standards that underlie the fine arts and basic sciences are based upon principle. We earlier recognized the principle of self-formulated, self-fulfilling rules (to which art often subscribes) and the dependence of science upon uncompromised methods. The artist and scientist try to say or do things optimally, whereas the basic canon of political life is compromise. Here expedience rules. What we later do with what we know, is a legitimate subject for ethical scrutiny and political will, but the search for truth is not.

If one is interested in the quality of light passing through parchment lampshades made of human skin, that is not innocuous unless one is willing to use one's own skin.

It was the chronic predisposition of political will to impose upon matters of principle that led Plato to conclude that there was no place in politics for a man of conscience. The game of 'truth and consequences' should be played in good conscience.

Off-Limits

We can infer from the above that there are topics that demand political intervention, those that require less resolution, those that may under certain circumstances and those that cannot flourish under political duress. There is another class of experience, solely the domain of private persons, which should be held to be politically off-limits on moral grounds. This class includes the overt interference with bodily and mental functions against the will of the subject. Unwarranted medical intrusions and restraints on belief (including religious precepts) fall into this category.

The acid test as to whether political activity is appropriate requires assessment as to whether any critically adverse effects upon persons could flow as a direct consequence of a particular activity. Political power can only be morally justified if scrupulously applied in the public interest.

Enduring Enigmas

The conflict of interests between individual convictions and the necessity to compromise in support of broader interests speak to the inherent difficulties of conducting political activity honourably and with good conscience. Conflicting interests are at the core of political activity because that is what politics are engaged in addressing.

It is the purpose of politics as a branch of ethics to be concerned with matters of the state or society as a whole rather than with the needs of individuals. Politics is dialogue. To illustrate this point one only needs to consider the absurdity of political monologue! Yet all that have interest and experience in the political field know political monologists that haven't yet clearly distinguished between their own personal agendas and the political business on hand.

Politics is the struggle to generalize the ethos of people in order to define the ethos of 'a' people. It cannot be overlooked that the state and society are pluralities of individuals. The political ideal of individualism functions like a protective 'alter ego' to hold at bay social impositions upon the tender ego. Individualism is not the political property of the left or right of the political centre; it is the natural stance from which personal interests are defended.

Politics has a forked tongue. When used in good conscience to promote social wellbeing, political life is a high calling. When used to foster personal gain, beware. Frequently these two are intertwined and it is this tangle that gives politics its bad reputation.

The limitations of politics are discernible in Friedrich Nietzsche's conclusions, not simply that human relations are reducible to a drive for creative power but, more significantly, that failure to perfect ideals drives a politician to seek retribution against others in order that he may maintain a position of strong political influence.

So-called political correctness is nonsensical. Politics has little to do with correctness and much to do with expedience. It is expedient to conflict mentally and verbally rather than physically. The morally correct measure of politics is to optimize in favour of the greatest good for the greatest number of people, subject to non-interference with individual rights and privacy, but only if this is consistent with one's intentions would such a course be politically correct.

Politics is like the tide; it is very difficult to get emotionally involved in a subject when you know that, on balance, the average state of human affairs is mean, and that all extremities of position will be resolved in time – unless one feeds on the contents of tidal pools at low tide, or upon beach-bound life thrown up at high tide. The smartest position seems to be to eat from whatever pool is available. In the meantime, lunch at mean tide doesn't seem too mean!

Decision-making to the left and right of the political centre is fundamentally unstable. While the goals of the political left and right may be to unify in their self-defined ways, their means divide. An analogy may be seen in observing relations between the great religions. If peaceful coexistence is the ultimate goal, the expedient standard that the great religions will necessarily require to adopt will be compromise, the very antithesis of what each espouses to be correct.

Citizens of a democracy who truly value their freedoms of thought and action have a moral obligation to resist the rise to power of supremacist elements, whether they be founded upon the basis of race, religion, aristocracy, wealth, or any other prejudicial ideological persuasion. If the government doesn't trust the citizens, how can the citizens be expected to trust the government?

The hinge upon which security turns suggests the need for a general understanding and the acceptance of the principles of democracy that derives its authority from the institution and maintenance of a flourishing system of liberal education, upon which the idea of freedom depends. We should take our cues from those of sage intellect, not from demagogues, ruffians or common fools. The future of democracy depends upon the general cultivation of elevated levels of intelligent thought and dialogue!

Edmund Burke seems to have got politics about right, contending that a statesman is one who combines a disposition to preserve with the ability to reform. The geyser 'Old Faithful' in Yellowstone Park delightfully illustrates the appearance of both novelty and continuity in its commanding performance. Upon observation it comprises elementary liquids, solids and gasses that dance interminably with variation in time and form.

All that remains for politicians to do is to find appropriate answers to the questions 'what', 'where', 'why', 'when' and 'how'!

131 Advice: Counsel for Consideration

We recognized in discussing education that the great predominance of knowledge has been accrued in the past and that much of such knowledge is secured in forms that are readily accessible today.

That such knowledge has, to a degree, been inculcated into the living, to do with as it may, is also recognized. In this sense the living are natural extensions of the past.

What is distinctively unique for the living is the set of circumstances that suggest and in large measure determine how such knowledge gets applied. The means by which a society interfaces first-hand experience with second-hand knowledge, and the burdens of prejudice that attend, are clear signs of where that society stands in relation to others with respect to advancing the cause of civilization.

Referrer

In the context of making judgements, those that have greater knowledge and experience may be expected to have more to offer than the relative novice. The novice legislator knows well to keep his mouth closed and his eyes and ears open until he has assimilated the rules of the game. To these ends he is better disposed to ask questions than to attempt to answer them.

Knowledge of context lends nuance to decision-making and to what ultimately determines the applicability of particular knowledge to particular ends.

Referent

Seniority comes with age and experience. While precedence, like history, is an invaluable resource upon which to draw, experience reminds us that there is no automatic default that guarantees repetition of events under similar circumstances! Political contexts tend to be complex and fluid, realities that experience and inexperience bring different qualities to bear upon.

Experience tends to default to precedence while inexperience invites novelty to a fault. How much of each is pertinent to any given situation is a function of political judgement. How astute Burke's observation was regarding the necessity for a balanced relation between preservation and reform, and how applicable this notion of the salient qualities of a statesman is to those of legislatures generally!

Reference

Ordering and reordering is what politicians tend to do, often not understanding the irony of a process of reordering that denies return to a prior order. Herein lies the fallacy of entertaining entrenched beliefs whereby the notion of conservatism is inappropriately applied to convey the concept of inertia as that condition most conducive to rewarding the good with the good life.

The burden of reducing the undue influence of prejudice rests equally upon the experienced and the novice, understanding that, while experience offers valuable insights into impending dangers, it is the relatively more open mind of the novice that enjoys greater flexibility in addressing novel conditions that arise within the context of a dynamic, ever-changing environment.

132 Law: Mankind's Ordination System

Insofar as experience is illogical, it is incomprehensible. Relations define the connections between existing entities. With respect to human experience generally, and personal relations in particular, we feel a compelling need to account for relations and to attribute causes. As we have earlier recognized, like causes beget like effects. From this understanding we may construe that if we can discover principles or construct rules of behaviour for causes, we may anticipate the probable effects and apply the rules selectively to our advantage. The function of law is to logically define and uniformly implement socially justifiable relations. It follows from this premise that laws must, of necessity, satisfy these parameters. The single point of contention with respect to law is therefore whether a particular law is correct, i.e. just, and if not, how it may be brought into conformity with respect to what constitutes correctness.

Absence of Law

Absence of law is anathema to the human mind. The trauma of the emergence of consciousness with its attendant ambivalence and stress taught us that our mental salvation depended upon a semblance of comprehensible order. Rules of order required to be invented in order to establish such order.

Natural Law

Nature has no laws per se; it operates according to principles, the most compelling of which is its impeccable resolution of conflicting forces. What we call natural law is a body of knowledge formulated by man to account for our observations regarding how nature operates.

The advent of life upon earth did not change or usurp these then-unwritten rules, nor did the subsequent evolution of conscious man. Conscious man simply raised himself to consider, in the light of his experience, what the natural causes and consequences of actions were, and were likely to be, and to share those considerations with his loved ones and friends, first by word of mouth, later recording them through writing. Why? So that he may place himself and his kin in the best possible position to ward off natural but unwelcome experiences. His instincts were self-preservative, just as they had been before his consciousness, much as other animals' instincts are today, but enhanced by newfound abilities to share experiences with others.

Nature is too vast, too complex and too changeable for man to comprehend in its totality. This actuality is more significant to the formation of laws than whether the laws themselves correctly correspond to what is actually happening in nature. Laws are bite-sized packages of understanding. Even when we believe that we have distilled the essence of a natural phenomenon, sooner or later we discover anomalies in natural behaviour that except, deviate or have been omitted from the laws as we have so defined them.

The 'laws' of nature classify natural behaviour by patterns of performance to enable us to address aspects of our environment selectively. It is the selective limitation of our fields of vision for the purposes of identifying common traits that determines the forms that we endow our laws with. Thus, the laws of nature are not the way they are because they describe the way nature is, but rather because they exhibit the way that we relate to nature. We chop it up and package it.

It is not that laws require to be impeccable, but rather that they should give us a reasonable basis upon which to understand our environment, particularly with respect to our abilities to predict and control future events. In time, in our pursuit of truth and our passion for excellence, we become more critical and demanding in the formulation of so-called natural laws. A case in point is the number of separate and differing determinations (more than twenty) made within a 200-year period regarding the true velocity of light. In fact, there are very few so-called laws of nature (science) that have withstood the tests of that period of time that still stand today.

Science claims to have deduced that all the natural laws that govern how the universe operates are embodied in no more than perhaps ten or fifteen principles of physics. If so, surely the governance of mankind is not as complicated and could be accommodated by a lesser number.

Manufactured Law

Laws are, by nature, manufactured. What is meant by this is that nature plays a role, but through man. Man presumes to define what nature is doing. Man attempts to perceive and articulate those principles as rules. Man draws comfort from his rules. His natural environment assumes meaning through laws. Laws are communicable and facilitate shared understanding, as a result of which man constructs ties of trust with his fellow men.

While it is increasingly useful to expand the body of knowledge of man's environment, and to represent that knowledge in the form of laws, the environmental phenomenon that is closest to him physically and emotionally, and for which he can discern no simple system of relational rules, is the society of mankind. He therefore feels the necessity to 'invent' behavioural rules of conduct that support his continuing need for security.

The enforcement of such rules in the name of security divides men into two camps: those that identify with, accept and accede to such rules and those that do not. We can already recognize that the existence and enforcement of particular rules begins to have effects upon society that are contrary to those intended, serving to divide rather than unite society; to promote insecurity rather than security. This places man in an equivocal position: either he must develop the ability to restrict and govern the behaviour of dissidents, by force if necessary (the conservative posture), or he must modify the rules in a manner that will overcome their objections (the progressive position).

The opposing doctrines of law, strict compliance versus interpretation of intent, are the drivers of legal, judicial and administrative controversy. The business of addressing these alternative courses is the profession and practice of politics, the essence of which, as we have seen, is the continuing competition between maintenance and reform. Laws are always fair targets for sceptics because history always bears them out to be time-sensitive and open to improvement. Law is the resort of the intolerant. The greater the degree of intolerance, the greater is the demand for new laws in the interest of establishing ostensible order. The fact that we manufacture laws attests to our incapacity to comprehend and beneficially relate to complex changing environments without narrowing our fields of reference. Our lack of tolerance for insecurity dictates the superimposition of order over appearances of disorder. In that sense laws serve the same function as beliefs in buttressing our confidence in light of our limited understanding of our environment.

Most laws have a bitter-sweet flavour; they offer the comfort of predictability on the one hand while denying flexibility and resisting change due to natural processes on the other. We impose structure upon amorphous appearances. Fear is the root cause of demands for law and order; fear of the unknown and the unknowable. We seem to have a copious incapacity to accept the ever-changing complexion of nature, of which man is but a small part. Laws are the products of our affirmative actions against her. At issue in any discussion of the role of law and order is the degree to which behaviour should be regulated and formulaic. The intent of law is to preclude negative possibilities, the effect of law is to constrain exceptions and the flaw in law is that it inhibits exceptionally rewarding possibilities.

The Legal Environment

Much as the laws of nature are too complex and changeable to comprehend in their totality, so man's legal environment has become too complex. By reason of this complexity lawyers necessarily specialize in branches of the law that befit their own personal dispositions, their focus upon the legal and moral niceties outside the areas of their special interests being diminished proportionately.

Reflecting upon our brief encounter with the Uncertainty Principle, we are led to conclude that the more we focus upon a single interest, the less we are aware of our environment. This is acceptable for short periods of time in order to ascertain the nature of an object of interest, but if we persist in engaging the narrow view we suffer a corresponding loss of attention to other aspects of our environment, including our higher obligations to society.

Many who administer law are extremists. For them enforcing compliance is undertaken with religious fervency. This should not come as a surprise since pious religiosity exhibits the same behavioural qualities in adherents who treat those that don't share their beliefs with the same intolerant disdain. Religious beliefs, while not natural in the sense of being scientifically accountable, are not normally legislated but are nevertheless manufactured and often take the form of rules of permissible conduct.

The Profession of Law

The legal profession, concerned with the generation, administration and interpretation of law, is unique in certain important respects. Since laws define social obligations and taboos that bear upon the lives of all citizens, the profession that specializes in the management of law is placed in a position of extraordinary responsibility for facilitating social opportunities and forestalling social failures. The word 'social' is used here purposefully to underscore the fact that law bears, or should bear, equally upon all subdivisions of society. The obligation of the legal professional is therefore to serve society fairly within the precepts of existing law, for which he bears a responsibility founded in trust. This fiduciary responsibility to act in the interests of the general good places the lawyer in the difficult position of reconciling the interests of his clients, constituents and personal convictions in correct and harmonious relations with those of society. Therein lie the trials and tribulations of the political animal.

The legal profession is indoctrinated to believe that no one is above the law. They cherish this idea as the pivotal point over which they break the backs of others. But let us take a look at this professed axiom of social stability and consider whether it is, first, historically enduring, and second, as fundamental as believed.

When we look at the laws of science we discover that almost no precepts held 200 years ago still hold today. The same attrition rate applies to legislation. There are about three times as many amendments to the US Constitution as there were original Articles. At first blush it appears that neither the scientists nor the legislators have a good grasp of their subjects. Upon closer examination they can be seen to be aiming at moving targets. Progress marches on and with it perceptions of what is what, why and how things ought to be. In other words the law is unstable. It serves man in timely ways. Law is there to compromise perceptions of unruly progress. Laws are revered as instruments of compromise by those who make and administer them. As long as compromise is the modus operandum, what hope is there for hope, what chance has Utopia? Where is space for optimism (positive thought) except through compromising compromise in the interests of optimization (i.e. making the best of)?

The Rule of Law

The expression 'rule of law' is a tautology. The nouns 'rule' and 'law' convey the same meaning. The premise behind the notion of 'rule of law' is that people need to know the commonly acceptable standards by which they can expect to relate to each other without incurring misunderstanding or negative reactions.

Political rule number one is 'Do Not'. Eight of the Ten Commandments in the Old Testament are negative decrees. Prohibitions are primarily intended to secure the authority of the institutions of church and state that issue them, and incidentally to afford greater security for their members. The extent to which people trust the institutions upon which they depend depends upon the degree to which the decrees coincide with the interests of the members.

The dictate 'Do Not' prevails today as the paramount message embodied in modern law. Little wonder that those who treat with law and order are so reticent about entertaining positive change. They are predisposed to saying no. By applying the mathematical 'rule' whereby the product of two negative values (that infer negation) resolve as positive values, the surest way to slip a piece of deception or legislative reform through is to couch it in the language of the double negative.

The legal profession is confrontational. Its modus operandi is the exploitation of conflicting interests. Just as a politician loses credibility and support as he moves further away from mainstream values, so lawyers lose credibility in the courts when they misrepresent a case through exaggeration, omission or deceit. Yet in order to gain the necessary leverage to out weigh their competition, they distance themselves as far as possible from the positions represented by their adversaries. In so doing they are drawn closer and closer to the brink of misrepresenting the facts at the expense of undermining those special fiduciary responsibilities. So here we have politicians and lawyers habitually pushing the envelopes of propriety in pursuit of conquest, with corresponding losses of commitment to the search for truth and justice.

When is much exaggeration, omission or deceit too much? Upon patent breaches of the law? And who makes, administers and interprets the law, and for whose benefit? So we have this iniquitous flying circus in which those that most fervently assert the dictum that no one is above the law are often the same that most furtively conduct themselves as if this was not so.

Natural Hierarchies

The laws of nature subordinate the laws of man. The 'Ten Commandments' of physics trump the 'Ten Commandments' of metaphysics. Where they are in conflict, the higher authority and the higher moral position is that consistent with the laws of nature. This position needs a little explanation since, as we have asserted, both are manufactured. If, to take an example, it was legislatively enacted that passenger airlines were to be held responsible for all damages arising from the failure of an aircraft in flight, then there is no inconsistency between causes and effects. If, however, it was decreed that aircraft shall not fail in flight, then such a decree fails to take into account the complexity of relations that enable flight and the probability that sometime, somewhere, a failure in flight will occur. Nor does such a decree take into account that gravity will ensure that grave consequences will ensue. Thus, in this case, the natural laws of probability and gravity override legislative or administrative laws of desirability.

Generally speaking, legislation is less rationally based than the laws of science. The laws of nature are distilled through observation, experiment and deduction. The laws of society are, by comparison, crudely constructed of compromise and often created in defiance of nature.

When science tells us that DDT will kill insects that reduce the yields of our crops, we advocate the use of DDT. When at a later date science discovers that in time and in sufficient quantities DDT will kill everything else, then we legislate that it shall no longer be used. So what nature does causes us to adjust our behaviour to bring our actions back into consonance with nature. Legislation, being predominantly negative, is reactive rather than proactive, intended to deny repetition of negative experiences. The same misunderstandings with respect to the aspirations of all men to be free apply to the acts of men subjugating others. Such acts are not simply inappropriate; they are misappropriations of nature's prerogatives.

Natural laws are analogous to nature's constitution by which she governs the universe. The laws of mankind are legislated by man for the benefit of man. Not surprisingly on occasion our best legislative efforts come into conflict with the natural constitution, at which times we are forced to reconsider them in relation to the greater authority and to revise them in the interest of forestalling detrimental consequences. From this understanding our laws cannot be regarded as sacrosanct and, logically, we should cherish those who, through the agencies of scepticism, imagination and initiative offer prospects of just adjustments.

Conflicts of Interest

All laws are always open to question. Thus, the environments in which they are accorded legitimacy, administered and interpreted are necessarily contentious. The adversarial legal profession, trained and experienced in combative ways, stands by like a reserve army of mercenary soldiers ready to take up any cause for a price. The fact that a lawyer's pecuniary interests and professional reputation are inexorably tied to winning exposes him to continuing suspicions of harbouring conflicts of interest. The motivations of lawyers should therefore always be held in question. The dictum 'question authority' is well founded.

Lawyers, by their special training and experience, are well positioned to compete successfully for elective offices and their interests guide them into the legislative branches of government from whence they expand the web of legal complexity, wilfully or by the nature of cumulative law that works to their benefit.

The legal environment in the US has become very complex, compounded by growing numbers in that profession quite out of proportion to their distribution in other societies. In consequence intense rivalry for business opportunities exists. Legal professionals are blessed with the ability to generate work for themselves by challenging the status quo, often precipitating crises requiring legal mitigation services and coercing others into paying for such services if need be.

The key to understanding why we proceed so slowly forward with social and judicial reform may be found in the format with which we address conflict resolution. Advocacy representation brings out the worst qualities of the advocates. Like politics, the legal process relies upon leverage. The further that an advocate can extend his arguments in directions opposed to those of his antagonist, the more force he can bring to bear upon the arbiters to decide in his favour. The potential rewards and the temptations to exaggerate, deceive and lie are enormous in a winner-takes-all competition. Winning at all costs becomes the operational imperative. The subordinate rule of litigation has become 'whatever it takes' (to win).

Legal process is the antithesis of creative thought. It tiptoes across streams of uncertainty on stepping-stones laid in the past. If one has crossed by a certain route before, it argues, this is probably a safe way to cross again. Always underlying such assumptions is the oft-unstated question: is there a better way? Precedence precludes progress.

The fact of precedence has become the justification upon which subsequent decisions are rested irrespective of the separate and distinctive merits of new cases, based upon the presumption that earlier cases have established the merits of the arguments in favour of a particular decision. Relying upon precedence to the extent that we do is tantamount to constitutional infortitude since in all matters there are changing circumstances and room for improvements in the interests of truth, justice and greater economy.

Legal Beagles

The principle focus of the modern-day lawyer is dominion over adversity. Lawyers generate and exacerbate adverse relations as a lifestyle. They zealously divide (or subtract) to conquer. Lawyers are negators. They have a predisposition towards distrusting others. That is why they are generally held in low esteem, even by those who profit from their activities. One has the sense that they are always looking over their shoulders for the stealthy antagonist. Lawyers fortify themselves with aspirations to wield economic and political power to offset their mental insecurity and society compensates them generously to maintain incentive to perform necessary but unpopular perfunctory tasks. One seldom meets a truly happy lawyer. As with all generalizations, there are admirable exceptions that one would hope any able young lawyer would aspire to become.

Only secondarily is the lawyer interested in compliance with the law, since laws are open to interpretation and application in support of vested interests. By the nature of legislative procedure laws assume forms that are conceptually compromised or intentionally equivocal.

Outlawyers

Like materials in testing laboratories, laws are compressed, stretched and bent by lawyers to establish the limits of their viability and their applicability to serve specific purposes. The modern lawyer has no vested interest in either truth or justice per se unless such virtues coincide with his client's interests or the dictates of his conscience. He is not interested in the resolution of conflicts except on his own terms.

In an ideal world any lawyer who failed to tell the truth, stretched the truth, gave false testimony or displayed arrogance or aggression would be held to account, subjected to disciplinary action and punished, since he would be breaching his fiduciary pledge. In a democracy the ultimate authority rests with the people. Ordinary people respect and observe the rules. It is the destiny of the extraordinary to seek improvements in them.

In Praise of Law

The laws of the land and the order that results in large measure determine our personal conduct and the semblance of order in our minds. Laws place limits on our multiple-choice options. We can rely upon certain uncertainties being resolved in certain ways regardless of the extent to which justice or injustice are embodied in particular laws. Laws give us reasonable expectations.

The human mind is in a constant state of impression. Laws limit choice, disabling man from responding optimally to changing circumstances. From the standpoint of the individual, laws in general tend to negate personal opportunities for survival and betterment. From the viewpoint of society, laws are effective in that they protect mankind from itself. It is man's proclivity for doing unto others what he would not have them do unto him that is his most socially destructive affliction.

In attempting to reconcile individual and social interests laws may be seen to be inimical but necessary. They are evil because they discriminate against men; they are necessary because in their absence man would discriminate against men even more. From this understanding we can construe the principle that the less that laws are necessary, then the more civilized or uncivilized man is, graphically illustrated by another bell curve diagram. The greater the number of laws on the books, the greater the distance that society stands from being truly free. Laws represent our best intentions to advance societies towards a utopian civilization while assuring that we will not reach it as long as we promulgate laws. An anarchic society represented by the jungle requires no laws, nor does an idyllic society represented by heaven. Only societies in a state of transition between the two require to be constrained from self-flagellation. Ninety million lawsuits are filed in the United States each year, one every three seconds! Simply stated: if we are not ready for that to which we aspire, we are not yet deserving of it.

The total mass of law signals where we stand in our passage from nowhere to you-know-where. At such time as mankind elevates itself to operate on principles, it too, like nature, shall require no laws.

133 Technology: A Driver of Change

In one year, 60 A.D., Queen Boadicea massacred 60,000 Romans and their vassals in revenge for their treatment of her family upon her husband's death. The introduction of the machine gun in World War I enabled the German army to kill 60,000 British infantrymen in a single day, a feat reduced to one minute with the dropping of the first atomic bomb on Hiroshima in 1945.

From these ignoble statistics we can quickly recognize that, as was the case for science, it is not the development of new technologies per se that is questionable, but rather the purposes to which new knowledge is applied – with unquestionable exceptions: a nuclear bomb is a nuclear bomb!

The history of technology is the story of man's escalating capacity to exploit energy, exert force, do work or produce change. That is the crux of his burgeoning and potential power.

Man's native ability to exert force is limited by his physique and condition. Lifting one pound through one foot every second is laborious. This harsh reality appeared to place limits on what was possible, but was this really so?

Imagination

What prompted the finches of the Galapagos Islands to utilize tools? Perhaps an accidentally well-placed twig awaited the force of a beak to pry for a gratuitous meal, but a single meal does not survival spell. A precedent may have been established but it took the stress of impending starvation and a modicum of memory to repeat the performance. Necessity provided the motivating stress, desperation the stimulus to break a routine and perseverance continuity in the flow of rewards. Imagination is not just resident and awaiting command, it's value is contingent upon use and development.

Understanding

It was interest in survival, prevalence over adversity and the appreciation of advantages gained through changes in behaviour that slowly laid the foundations to man's understanding that he had within his power the latent ability to improve upon his lot. The prying of mind into matter began to deliver rewards beyond his imagination. The application of principles came to be understood as the determining factor governing man's ability to ratchet up his performance to higher levels of productivity. Let us examine a few simple examples to show how technology evolved and compounded advantage.

Leverage

The application of a pry bar to raise a load drew upon the principle of leverage whereby, through increasing the distance of a force applied over a fulcrum or pivot point, a proportionate decrease in effort was required. This ratio came to be known as the mechanical advantage.

The roller, later refined as the wheel, is a hybrid form of leverage whereby the spokes of the wheel perform the function of multiple lever arms. Similarly, a pulley is a special application of the leverage principle turning about a fulcrum or axle that transfers the load through continuous flexible lines.

An inclined plane distributes a load between a ramp and the effort-force proportionate to the angle of the incline, taking into account the coefficient of friction. A wedge is simply two inclined planes back-to-back that move in relation to a load that is stationary. From this understanding screws were developed as continuous wedges wound around a cylinder and rotated on an axis. In this case the wedge or plane describes a circular or rather a 'helical' path about an axis rather than a straight line.

Compound Leverage

Most mechanical devices (machines) combine two or more principles for the primary purpose of enhancing the mechanical advantage. Early mechanics dealt with relations of materials, not with changes in the state of materials exemplified by chemical reaction. Subsequently fluids, gasses and chemical relations were found to offer great potential for amplifying mechanical advantage.

Machines are conservative by nature. Their principle function is to conserve whatever resources are in short supply and to amplify the utility value of that commodity by design. The greater the ratio of output to input, the greater the efficiency.

Engineering is exemplary in the sense that it is wholly dependent upon the science of mathematics. Rational designs are the only ones that work – the more so, the more so!

The invention of tools solely for the purpose of making other tools constituted a quantum step in the evolutionary development of technology.

In the systematic application of scientific knowledge, technology's most enduring purpose has been to further extend security. Not surprisingly, many early devices were developed as mechanisms of defence. The defence industries continue to develop machines at the cutting edge of technological know-how intended to secure time, space, energy, matter, relations, the intellect, volition and the affections against perceived or real threats to society.

Efficiency

Efficiency is measured as the product of output divided by input. Whether expressed explicitly as power-to-weight ratios, miles per gallon of fuel, ratios of labour, materials or energy invested in a product or in monetary terms, gain is the principle motivation underlying technological development. Efficacy is the power to produce an intended effect, efficiency its measure of success. Short-term interests have prompted men to improve upon their native productivity while the cumulative effects of short-term gains describe the long-term effects. Expediency is the nub of expedition, the driving force behind technology.

Less is More

Mechanics imports dynamics: the movement of bodies in response to forces. What was so ingratiating about early mechanical design was that it lent itself to pre-testing in theory through utilizing the charms of graphic proportionality inherent in geometry.

Our prying minds and ingenuity have led us from the standard platform of seeing as the basis for believing, to believing of necessity as a consequence of not seeing. We have run the whole gamut from huge clanking mechanical technologies through the electromechanical transition to miniaturized electronics, a transformation that renders our sight inadequate to the task of verifying the functional integrity and performance of our labours as we extend our division and subtraction into diminutive scales. No longer may we rely upon geometry as the confirmatory analogue. Hello to robotics manufacturing, nanotechnology and cybernetics, the development of man-made mechanisms that behave like goal-directed living things.

The Age of Engineering

A technical system signifies the pre-ordering of technical elements. It was this ordering, in the interests of efficiency, that governed the direction that engineering was to take from the beginning of the industrial revolution.

The time-tested methods of trial and error came under greater scrutiny and control as the scientific method gained greater acceptance. Materials testing and design became complementary activities while numerical calculation supplanted empirical design. Tasks were learned and then performed repetitively by rote and in-line, from whence assembly lines evolved.

The quest was on for improvements of any and all aspects of human industry. The concept of engineering, i.e. deployment without excess, began to be applied to non-mechanical fields of work and the term 'engineering' was appropriated and prefixed with modifiers like 'cost', 'market', 'social', etc. to denote the fields that had adopted its methods.

Automation

The logical consequences for progressive cultures that replace human labour and judgement with technological work and controls are automation and redundancy. These two by-products of technological development appear on first acquaintance to be as cause is to effect, automation out-performing hand productivity in quantity, quality and cost in a measured time frame, while redundancy denies the traditional demands for human labour and the livelihoods that flow from work. The proponents of automation argue, however, that it serves the needs of greater numbers at lesser costs, frees man from the shackles of monotonous repetitive tasks and creates new industries and new opportunities. That, they insist, is progress.

Progress

Progress implies movement and improvement. What we have just observed may qualify as progress but the term 'qualify' requires to be qualified itself on two counts: to ascertain the direction of movement with respect to a particular claim of progress, i.e. whether it is positive, neutral or negative in its impact; and to whom an improvement is relative? The traditional concept of heaven embodies belief that the deserving on earth shall be accommodated in a society where all will be elevated to higher standards of satisfaction in a life here'after'. This is the identical premise underlying technology in all respects except that the suffix 'after' is omitted.

Acceleration

Progressively less time is being applied to yield progressively more products. If Einstein got it right (i.e. greater acceleration translates to greater mass), then we cannot be progressing towards greater acceleration without exponentially greater effort. Underlying all these machinations, intriguing and seductively attractive as they may be, there must be something more profoundly satisfying than merely doing more and more with less and less. So what is the nature of the more towards which we aspire?

Economic Leverage

The accelerating flow of events offers possibilities for providing for burgeoning poor populations, good prospects for conserving natural resources and certainly works in the interests of the stealthy becoming wealthy; but is that all: mean servility at the bottom of the human pile and mean incivility at the top?

Technology is a physical manifestation of the application of economic principles, its evolution the process of piggyback leveraging to compound interests. Like law, technology endeavours to advance human interests, their differences being as between software and hardware. As interests compound they generate complexity. We break that complexity down into relatively simple packages.

Specialization divides authority, estranging one specialist from another. The division is one of values whereby one specialist optimizes his values in one currency while another does so in another. A common denominator is sought to enable comparisons of performance from which priorities can be determined. Money becomes that intermediary value system by default. So in order to do business with the rest of the world the lawyer needs to generate money, as does the technician; not for the purposes of continuing the good work, but to force issues of compromise favourably in the direction of their interests. Money now becomes the lever arm that levers all other lever arms, quite regardless of whether monetary value could ever be a suitable substitution for deeply felt human values.

Our resourcefulness has led us to invent devices that are more productive than ourselves. At first this efficiency ratio was of an order of 200 or 300 per cent, as the lever and pulley demonstrated. The industrial revolution accelerated the pace of accomplishment and the ratios of power increases have increased geometrically. Today we stand in awe of our inventions, relegate our workers to the status of tools and acquiesce to the dictates of mammoth enterprises that treat people as disposable industrial waste, unworthy of recycling at the company's expense. We have learned to revere the juggernauts of technology and pay homage to the deliverance of their raw power. Too often in practice Robert Browning's dictum 'Less is more' (from 'Andrea del Sarto') has been misappropriated and subverted, adding an unspoken parenthesis 'for the few'!

Private Leverage

The purposes of developing technology or engaging in almost any other human enterprise have become marginalized to the pursuit of money simply because money is the currency with which we agree to exchange values. The doctrinaire pursuit of money, and the power and privilege that it buys, has become the new religion. This is the floor to which our cultural values have settled.

This paradigm has been inculcated into the human psyche to the extent that we have become conditioned to seek and expect greater returns on less investment in our human relations. The cult of the profiteer is going to take a long time to disencumber and supplant with a more affectionate standard of human relations. An economic system that is predicated upon giving less and taking more impoverishes society in exponential proportion to the degree that it enriches private investors.

Technology has broken down, not in the sense that it no longer works, but in the sense that it no longer performs particular work as its primary purpose. It has become derailed, heisted, hijacked away from its potential for serving as the strategic engine of social benefaction to becoming the tool of choice for leveraging monetary advantage. The point is not that technology is unpopular or undesirable but that the populace does not control abuses. The policies that economic capitalism espouses are antisocial by self-definition. Ends are held to justify means. It is the assumption of arrogant pseudo-aristocratic airs in presumption of entitlement to privilege that estranges the capitalist from the common man, not his native abilities. The consolidation of private enterprises into ever-larger and more impersonal conglomerate holdings, ostensibly to appreciate greater economies of scale, is a flawed concept. Such leverage works for the few, for a while. The fewer people that share the profits, the more each investor gets in return. The manipulation of the many by the few has relied upon strategies of intimidation using law and order as a pretext to prop up the 'might is right' dictum. The Chicago gangster Al Capone was very adept at using this strategy, in his case using the law of the gun and order as command. Monopoly and extortion are logical outgrowths of restrictive marketing. If either means or ends are mean, they must be brought to an end by other means.

Having earlier established that acceleration in the interests of the few and at the expense of the many has practical limits, we must ask ourselves in what direction should leverage be applied in order to satisfy the dictates of both logic and the emotions?

Social Leverage

Personal integrity has something to do with balancing the needs of society with the needs of the individual, not promoting one at the expense of the other. Our integrity serves our collective affections, all the things that we hold dearest in our hierarchy of values that become etched into our constitutions.

Something that is conceived and operated by people who care about what they are doing is superior in critical ways to that which is produced impersonally. Conservatism in this regard is properly seen as the defendant of personal identification not only with the product but also with the process.

Underlying all technologies is the question whether an initiative is justifiable. The moral question is inextricable from the modern processes of civilization.

At first man only needed to account to himself, later to his family, community, state, nation or race in ever-increasing scales of magnitude. Now we must ask ourselves not merely who benefits or should benefit, but who pays and in what ways; and what controls and how? As long as the 'Peking Butterfly' has flown, we have all known that all is related to all else.

In order to deliver upon its original promise, technology must be reapplied to the deliverance of its greatest potential: the greatest benefits for the greatest number. The urge to improve upon the human condition is fundamental to humanity.

About Turn

Second millennium machines are analogous to animals. They do the things that animals do but with more (or sometimes less) efficiency. We even rate their potential for performing work in horsepower. They burn hydrocarbons (organic compounds composed only of hydrogen and carbon) much as we consume carbohydrates (compounds comprising carbon, hydrogen and oxygen), excreting carbon dioxide (CO^2) and other waste products much as we do. The scale of such mechanical activity has compounded exponentially to the point where the combined material consumption and waste output of growing populations and their machines far exceeds the ability of nature to counteract its effects through plant metabolism. We are exacerbating a deficit economy fully dependent upon depleting resources! Logic driven by necessity dictates that we reverse that process whereby we build machines that are better analogized as plants.

Contemplate power plants that truly emulate the alternating functions and performance of plants. The fuel of plants is radiant solar energy. Solar energy is applied to partition water into its two constituents, hydrogen and oxygen. The natural plant respires carbon dioxide. The hydrogen of water is fused to the carbon of carbon dioxide to form organic carbohydrates, through which process excess oxygen is expelled as waste, the effects of which on the one hand is to consume surplus carbon dioxide while on the other hand generating additional free oxygen. The deficit is drawn down by the first process and offset by the second. Time for another analogy.

Imagine a steam locomotive that consumes large quantities of carbon dioxide and other noxious gasses through its funnel, fuses the carbon component of carbon dioxide to the hydrogen of water through the auspices of solar energy, drives the train (backwards as it were), piles up fuel in the form of hydrocarbons in the tender, replenishes the boiler with water, and emits large quantities of pure oxygen. Today just such a technology exists and is in limited production. Tomorrow we will be so familiar with this plant simulation process that we will have such a plant, known as a hydrogen fuel cell, on every rooftop, or a larger, more efficient equivalent in every neighbourhood.

A simplistic way of looking at such operations is to recognize that if a process is reversed, say a stream of water or electrical energy, the turning effect that such a reversal will have upon a turbine will not change except to change the direction of drive. If you want to drive in a certain direction (say in your car), you do not set off in the opposite direction.

The machines of the third millennium will automatically assess imbalances in nature and convert to operations that serve to rectify such imbalances, as forward or reverse operations of the locomotive analogy. This may sound a little 'loco' but with the right 'motivation' – all ends well that is well.

Reflecting upon our thoughts concerning 'will', and specifically the inanimate will to 'want to be', technology is leading us in the direction whereby we can attach 'intelligence quotients' to familiar objects and systems that enable them to 'talk to us' concerning their conditions and needs as exemplified by the emerging technology of **r**adio-**f**requency **id**entification. RFID uses tags and readers to track and communicate the 'whats, wheres, whys, whens and hows' of tangible things.

The lesson to be learned is that we cannot afford to take our inspiration from prejudicial preferences for past styles of behaviour abstracted from eras when mankind was less aware of the consequences of his actions than he is today. We must turn to another source of inspiration, nature; a course that upon occasion we have unwittingly profited from in the past.

When innovative technologies pre-empt private greed, the opportunities for economic abuse dissolve. The ultimate determinant as to whether a venture is undertaken or not should surely be whether it is worth doing in its own right.

We close our brief discussion of technology with the reflection that the nineteenth century is remembered for the harnessing of raw physical power, by which man rent himself from his pedestrian nature to gain great material wealth, but lost his virtue of humility. The twentieth century delivered to man the means of control, as the supreme arbiter of fate on earth, without the ability to distinguish between his goad-like propensities and his god-like potential. Logic and the emotions strongly suggest that if we are to exercise the ultimate control over our burgeoning technologies, the twenty-first century will necessarily require to be remembered for its contributions to the morality of doing.

134 Perplexity: The Child of Complexity

The word 'perplexity' conjures up a hybrid idea combining the limits of a 'per'-son's ability to cognize com-'plexity'. That is the impression that the word is intended to conjure.

Perplexity

Perplexity is a signal detected by others concerning the behaviour of one who cannot make meaningful connections between his experience and his understanding of reality. We seldom admit to perplexity because it suggests that we may not be as mentally dextrous as the company we keep. This perplexity is suggestive of a boundary condition between finding correspondence between disparate qualities and abandoning the search in favour of accepting a fantasy of multiple realities. We are all susceptible to perplexity. Most of us simply haven't crossed the threshold into permanent residency.

Autism

Those who are so disposed act out their fantasies auto-matically. Their lives turn about their auts or selves, not that others do not, but in such cases the turn of the mind is with seeming disregard for causes and effects, and past and future; being absorbed in need-satisfying or wish-fulfilling fantasy as a mechanism by which to escape from external realities. They have entered the private world of the autocosm, a self-centred microcosm. Such a disposition, when found in young children who cannot form meaningful interpersonal relationships, is known as Kanner's syndrome, believed by some to be a childhood form of schizophrenia. With regard to causes and effects, causes of autism are genetic, arise from chemically abnormal prenatal conditions or are due to stresses arising from external environmental circumstances that are beyond the capacity of the individual's mind to redress.

Schizophrenia

Chronic perplexity finds expression as schizophrenia. We have visited this subject at the birth of consciousness without using the S-word. We shall doubtless visit it again. The term 'phrenia', from the Greek *phren* meaning mind, refers to a disordered condition of mental functions. As used in the *Iliad* at a time when there was no word for consciousness or mental activity, the term more closely resembled the modern expression 'guts' from which we constitute our 'gut' feelings. Thus we can interpret schizophrenia as a relapse back into an Iliadic state of mindlessness when mental disorder was unknown and unknowable because everyone was schizophrenic! The extent to which heredity and environmental factors may contribute to schizophrenia is controversial, a condition that could be explained by the number and possible permutations of contributory factors. From a biological perspective mental disorders are chemical in nature; from the viewpoint of a chemist they are physical, and from a physicist's standpoint they are biological in the sense that they are governed by the idiosyncrasies of living organisms. Mental deficiency is literally a state of something 'missing' in an affected individual that is normally present in others. In general such absence translates physiologically to a smaller brain, considerations of age and sex being equal.

Mens Sana (Sound Mind)

The antidote for complexity is simplicity, the imposition of simplification as a means by which to continue to transpose experiences into meaning. The aphorism 'Keep it Simple, Simple' is not simply a path to better understanding; it is 'the' path to absolution from excessive mental stress. In this context it is more important for mental stability that a belief be adopted than that such belief be soundly based. The simpler (less cerebral) the subject is, the more subject he is to holding simple beliefs. Herein lie the limitations of *truth* as a utility! Those who are more animate than humane (i.e. are less evolved) are more inclined to operate in the automatic mode. Simply stated, Simple Simon relies upon his auto-response system to automatically check novelty for correspondence with his 'imag(es)-in-a(c)tion' and to reject all misfitting signals. From such realizations we are led to 'believe' that reality is in the mind of the beholder. Complexity is stimulative, simplicity tranquilizing. The belief system shields the sensitive soul from unwelcome impositions. The system *in toto* is represented by a summary piece that the subject shares with like-minded souls: doctrine. We can now understand the fixations that people, and groups of people have with regard to what is acceptable and what 'should' constitute reality, most conspicuously evident in unyielding political and religious attitudes embracing the acceptance of external authority without questioning. A 'phren' in need is a 'phren' in-deed!

135 Competition: The Struggle for Survival

Having observed the physics of simultaneity whereby no two things can occupy the same space at the same time; and having therefore resigned ourselves to providing for the orderly sequencing of events so that one event occurs before or after another at any particular place in a particular time frame, we thereby recognize that, in accordance with the rules of nature, a state of competition exists between like-minded and/or like-blinded things for occupancy of any particular space at any particular time. This competition is so fierce that nothing is absolved from the competitive impetus. Once we thought that we had found an exception, that nature abhorred a vacuum, only to discover the counter-force; that vacuum abhorred nature. Insofar as vacuum is constituted of nothing, nothing is in competition with anything and everything. 'Nothing', being an entity without identity, is exempt, and not exempt from participating in a competitive environment.

We have also travelled the path that showed us that all worldly matter competes for a straight-line path to the centre of the earth but is only successful if it precedes all else in succession, subject to a few niceties like relative specific gravity, temperature etc.

Animals Do It

The same simple rule applies to the living. All animals of like kind seek like food to sustain themselves under circumstances of natural privation or over-population. Enter Darwin and Wallace to explain what happens next. Those that don't have the competitive edge, because they are not strong enough or lucky, tend not to prevail.

Why Me?

Why does nature produce life in such great abundance? Is there not a great excess of waste? The answer depends upon attachment of purpose to creation. In the absence of evident will, we must presume purpose to attach to our personal interests and to include causes and effects. The cause and effect of the creation of quantities of new life far in excess of the numbers required to maintain balanced populations is evident in the means by which life grows and maintains itself over life spans:

I see the world the strangest venue
Everyone is on the menu
God is witness to the feat
That meat eats meat eats meat eats meat

Fish eat smaller fishes
Eating smaller fishes eating
Smaller fishes eating smaller
Fishes eating smaller fish.

So there we have it. Each life has but one death but is responsible for countless deaths if it is to survive and mature. There is no waste in nature. Everything is recycled.

The welfare of others is no pain to one's self, unless one entertains the squalid view that the impairment of others better befits oneself in the competition for more fortunate circumstances.

Why Us?

'Why us?' may appear to be a redundant question given the foregoing exposé. But mankind has evolved into something of a god-like creature in the sense that it has accrued the power to prevail over all other life forms and to put any or all of their existences at risk. So the question arises whether or not there is a special condition that applies to man whereby he is encumbered with the responsibility of ensuring that lesser forms of life are not extinguished due to his misunderstanding as to their roles in the overall scheme of nature, with particular respect to his own interests. It is, after all, only man's interests of which he is aware that will motivate him to act in such interests. It is therefore beholden upon him to become aware of his own interests to the maximum extent of his capacity. By 'he' we refer to mankind rather than to an individual, understanding that we are all mortal. We have, very slowly, acquired an interest in understanding this sobering sovereign responsibility. Inasmuch as each of us is capable of inflicting unknowable damage upon a well-tempered process of approximate balance between a myriad of competitive live forms, each has the power of a sovereign and each bears a share of that sovereign responsibility.

How Do You Do?

Being a god-like sovereign and assuming such responsibilities we are led to conclude that, by virtue of our greater faculties, we have greater obligations to take a higher view of the operations of nature than do the lesser flocks. But amongst our own flock we observe that there are lesser folk who do not recognize any obligation to take that higher view. So how do we do for others what they would not do for us, their own kind? The answer appears to be two-fold: first patiently tend the flock by setting an example, explaining the whats, whys, wheres, whens and hows concerning preserving natural variety as well as one understands them; and, if that does not persuade them to reform their careless characters, then invoke the example of nature and enter the arena of competition. As we have come to understand, the ultimate prize of competition is survival. Survival for what purpose we may ask? To which question the most persuasive answer has to be for the purpose of affording protection for the continuity of the survival of life in its fullest diversity of expression.

To survive we must prevail over adversity, therefore, those with the broadest, more universal concerns, under peril of extinction, must prevail over the unthinking to avert the unthinkable.

In so doing the only species capable of inflicting irreversible detriment upon the earth's ecosystem opts for the longer-term interests and higher responsibility of averting such affliction.

Is this an echo of the Illiadic man-god mind at work, albeit operating at a (slightly) higher level of evolution? Maybe, but we should remember the lesson of 'maybe' gained from our early thoughts concerning thought as the imaginary dialogue of the intellect; that 'maybe' as 'the Achilles' heel of (man's) intellectual existence, the ghost of his former traumatized state of mind, taunting him at every turn, goading him on to take chances and make mistakes'.

The lessons of those higher purposes of higher education require to be learned well and acted upon by all means. Let the caveat of a modern Illiadic war hang heavily over mankind as long as he is unkind to his kind and life-kind. Prevalence in competition is ultimately the responsibility of the responsible.

Life is problematic. Man has the cult of competition deeply ingrained in his psyche. As populations increase and difficulties compound and become more pervasive, the need for a competitive way to our competitive ways, for cooperation rather than competition, will become increasingly more apparent in order to reduce the problems of society.

Nature's secret force demands that forces resolve into a state of approximate equilibrium. In the affairs of man we can do this with good will, a smile and a handshake; or we can do so with denial and revolution.

136 Collaboration: Compound Interest

In Section 1 we contended that time was the ultimate parameter of life. We must now ask ourselves what additional enabling conditions govern the viability of life as we know it. We can presume that all the conditions that currently contribute to the diversity in life forms minimally pertain to their future prospects of survival. Insofar as evolution, aided by entropy, develops from simpler to more complex forms, conditions that would enable greater variety would assume greater complexity too. From this rationale we can extrapolate that, contrary to our current penchant for exploiting resources to the point of exhaustion, greater biodiversity demands that we further advance variety in opportunities for novel forms. That is to say, we should collectively co-opt resources for the purpose of generating greater variety. To that end collaboration is the natural and inevitable means.

Dark Portents

Man has learned how to exploit the earth to his short-term advantage. That process necessitated overcoming competition between opposing forces.

Our psyches have become imbued by the idea that competition is the sole means by which to achieve any and all ends. We have acquired the ability to overcome life itself in all its splendrous forms. That could occur in a single catastrophic event or more subtly in increments through our negligent failure to heed the warning signals of stress in the biosphere.

Our historical experience of competition, while operating at progressively larger scales, has been confined to competition between factions of humanity. Competitions have been played out in the field and the field has always been more or less defined. In terms of mortal combat, even the greatest battles have had a site, boundaries and no-play zones. Even in so-called world wars of the twentieth century, the greater part of humanity has been relatively insulated from the fray. The interests of the adversaries have been clear, but more significantly, the event of a win by either side ensured that the processes of living would continue much as they had before.

Today public enemies number 1, 2, 3, 4 and more are stealthy. They disdain containment, bound boundaries, deny political and geographic territorial prescription and, worst of all, compound their negative impacts exponentially. Without suggesting which is the more insidious at any particular place and time the 'bad guys club' has to include global warming, pollution, the indiscriminate spread of weapons, space war technologies, the depletion of finite resources, burgeoning populations, intractable diseases of epidemic proportions, biological hazards, abusive narcotic drug trafficking, species eradication, deforestation and so many, many more, subdivisions of any one of which is a greater challenge than man has ever confronted before. What is most significant and most disturbing is that we do not have either the mechanisms or the collective will to surmount these difficulties, no one of which can be contained or excluded by a single political entity. If we then add to these difficulties the causes of untimely premature death attributable to malignant and infectious diseases, and morbidity rates attributable to causes over which man has some theoretical control – accidents, poisoning and violence – we begin to appreciate the extent to which we are at risk to the impact of forces that cannot be addressed or overcome through the auspices of competitive behaviour. The one bright light at the end of the tunnel is that if negative impacts compound exponentially, then any remissions or reversals in their progress can also achieve exponentially effective results.

Help!

'Help! I need somebody, Help! Not just anybody, Help! You know I need someone, Help!'
The ballads of The Beatles resonated deeply with so many people all around the world in the latter half of the twentieth century, not simply because they were the genial music that they were, but

because they spoke so profoundly to the plight of each of us as individuals who at times feel the seemingly unbearable burden of facing the world alone.

'Oh yeah, I'll – tell you something, I think you'll understand. When I – say that something, I want to hold your hand.'

This is not just sloppy sentimentality; it speaks to the essence of the problem of coexistence between people, to needs and to the apparent absence of relief. Everywhere about us we witness unmitigated, relentless competition. Only at the extremities of our life experience, at times of extraordinary distress, in the face of natural and unnatural disasters, do we switch off our competitive urge and pitch in to be that 'somebody' without counting the cost.

Where is that 'something' when it is so often wanted or found wanting? What does it take to light a fire of hope every day in every heart?

'All you need is love, all you need is love, all you need is love, love; love is all you need.'

Something

Something, some force, seems to be required to catalyze our greatest abilities to perform our greatest deeds. As to whether it is fear, love or some other stimulus is for each of us to discover in his own way, but how can so many miss the point? How come so many take so long to get it? How is it that education does not teach a basic system of human relations precedent to teaching the tricks of advantaging? That would be a trick! That would be 'something'!

Compound Interest

The ultimate human interest is humane interest. That is the standard by which we all aspire to be treated as individuals. What is so hard about agreeing upon such a simple principle? The mass production of care is not inconsistent with logic. The mathematical function of progression whereby each increment of success in advancing a principal is achieved by multiplying it by additional interest to attain an ever-increasing stock of capital is a well-established economic precept. It 'works' whether princip<u>al</u> is spelled with an 'al' or an 'le', whether interest is fiscal, intellectual, emotional or some subtly balanced combination.

To do unto others as you would have them do unto you is not simply a Christian canon, it is good business and common sense. It is in sync with sound capitalist thought and, not by coincidence, also with the interests of society. If, in seeking beneficial relationships, one can secure the notion of willing giving and willing receiving in symbiotic reciprocation, it is hard to imagine a more satisfactory state of human relations.

Co-labor-ation

To labour together for common purposes we need to combine the special talents of each other in a way that we maintain the willingness to give and receive, committing to such behaviour as a cultural expectation and norm. If this sounds a little too utopian to be practical or possible, take a look at the behaviour of the humble bee or ant. The ant can carry in excess of its own bodily weight of supplies back to the community nest. The bee generates honey in excess of what it, its peers and its progeny need; excess enough to write off large quantities as losses to marauding scavengers like Pooh Bear and you there. There is no way that bees and ants could survive if they were in constant competition with each other. They have had to find another, more viable means of surviving in coexistence through cooperation. They made this quantum step in the interests of their own security. When we observe the animal world we can readily discover species that are substantially self-sufficient, the eagle is a good example; those that are entirely dependant upon their own kind for mutual assistance; and those that operate in modes that combine the two. Man has very slowly progressed from the former towards the latter state of being. He has the advantage of discernment and discrimination whereby he can assess which mode or degree of mix serves his specific purposes more effectively, but he cannot yet feel comfortable with the fact of his dual existence as an individual and as one of many. He still carries vestiges of his bicameral mind, each half of which pays homage to a different god.

Sharing is a sophisticated idea requiring a measure of personal detachment from the present in order to appreciate how much better future relations could be. So the three strongest arguments in favour of collaboration are: first, that it offers the best prospects for survival; second, that it affords the surest passage to a state of society whereby humanity (what we are) may advance to a state of 'humane-ity' (what we could be), 'humane-ity' simply being a higher standard of living than we are accustomed to; and, third, by virtue of its inherent capacity to compound interests mathematically, collaboration offers the fastest track to getting there. Are these not worthy goals?

137 Harmony: Conformity to Principles of Integration

A Prescription for Tranquility

Harmony arises out of combination or relation. Insofar as it is distinguishable from its antithesis, discordance, it may be equated with integrity whereby it manifests itself as the essence of wholes rather than of fragments.

The Pythagoreans sought explanations for their environment in terms of perfect order, conditions that were, if not changeless, endless and therefore inviolate. Such changes were most clearly evident in the night sky wherein patterns and relations appeared to be circumscribed by unending cyclical motion. In these they saw unity in variety which harmony they referred to as the Music of the Spheres.

In music, the simultaneous and consecutive sequencing of musical notes into chords renders it legible, legitimating its progression to subsequent chords. Thus, harmony is a system of structuring events that occur periodically in accordance with predetermined patterns.

Periodicity, an essential characteristic of harmony, is not confined to music, but, as we are discovering, is resident in everything that we reveal through science. The Music of the Spheres is discernible at the sub-atomic scale. Periodicity appears to be a vital constituent of the harmonic repertoire of nature.

The quality that defines harmony and makes it recognizable is that of correspondence, agreement or accord. Thus, for example, one may represent community relations as being harmonious when there is freedom from conflict.

People cannot be represented as being a community in the absence of a binding affinity, much as aggregates cannot be represented as being coherent in the absence of an adhering agent. Harmony is to community as cement is to concrete.

Maximum freedom from conflict within a particular context appears to occur when such agreement bestows tranquility upon elements, setting the stage for its continuance.

Harmony as an Aesthetic Quality

The essence of aesthetic harmony is pleasure, the absence of harsh contrast or contradiction. As to whether harmony is an active force or a passive condition is open to question but since it can be recognized to function as a cause as well as an effect the answer appears, like the definitions of time and space, to be circumscribed by circumstance. Perfect silence seems to qualify as being aesthetically harmonic since its effects are eminently tranquil. So, how does one recognize, better still distil, harmony?

What defines the critical threshold beyond which harmony is rendered mute? To what extent can we justifiably insist that harmony should or cannot be present? Is harmony incremental and measurable by degrees, or indefinite? Is the presence of harmony simply a function of subjectivity, to be sensed as a relative aesthetic value?

Calculable Harmony

If we try to reduce harmonic experience to mathematical terms we find that we can achieve a significant measure of success. For example, we can equate harmony with congruence in geometry, or with the relation of sine to cosine in trigonometry with a sense of absolute assurance. Similarly, we can establish musical notes as being synchronous if they bear certain mathematical relations in frequency, as between two Cs spaced one octave apart, where the upper note registers a doubling of the vibration frequency of the lower note. Thus, harmony may be found to be a useful means by which to represent proportion. The same set of frequency relations applies to any and all subdivisions of the electromagnetic spectrum. From the above we can discuss language characteristics that are common to all periodically recurrent signal systems. The discipline concerned with the measurement of such systems is known as Harmonic Analysis.

Good Vibes

We have inferred that an aesthetic appeal attaches to synchronous sounds that suggests a basis for our love affair with music and which in large measure prompts composers to compose.

Bad Vibes

Inimically, the same qualities that attract us by virtue of their recurrence also serve to detract from the integrity of man-made forms. The effect of similar sound waves concurrently impacting membranes at critical frequencies is to combine the forces of individual sounds into compound sounds that generate sufficient concentrations of force to destroy the integrity of material surfaces exposed to such combined forces. Metal fatigue is attributable to the breakdown of a molecular structure due to excessively extended cyclic loading. The ear diaphragm has similar tolerance limitations to sonic vibration.

In the ideal world we would hope to place critical limits on the incidence of bad vibes while promoting conditions that enable maximizing good vibes!

Harmony by Design

Energy in any form emanating from any number of sources has the capacity for being concentrated and converted into other forms the effects of which may be positive or negative in relation to mankind. It behoves man to search for means by which to convert negative vibes into positive ones, by which means he not only redirects such energy to good ends but also transforms himself into a state of harmony with his environment.

Whenever we identify conflict or competitive forces we need to look for a path to resolution that will nullify excessive force. Whether they are redirected, amplified, resisted or left to run their natural courses is a contextual question.

The question finds its authority in the primacy of the greater forces in the greater environment. We need to examine what the prevailing physical, social, political, military, legal and other forces are in order to establish what is possible (or impossible and can therefore be disregarded), always deferring to the greater interest, be it life, freedom or happiness, recognizing that might is not always right; being subject to will-as-probability and will-as-intent.

Harmony by Default

We should recall that to those early acquainted with the magical precision and invariable correctness of geometry there was an irresistible temptation to identify its mystical qualities through cosmology or to the only other semblance in their experience that garnered equal awe; Godliness. So much was this so that the one was irresistibly attributed as the province of the other. Arising from Plato's mysterious work *The Timaeus,* the natural extension of such a connection was to interpret that the whole universe comprised a form of measurable harmony. As such, harmony is an expression of natural, some may say providential order, which we can deduce incorporated moral order.

However, in order to gain understanding sufficient to distil the essence of harmony we are drawn to hypothesize an unlikely elixir comprising equal parts of precision (mathematics) and misapprehension (mysticism). To many this should not come as a great surprise since we inhabit a world in which much that we do much of the time is to operate in a blur somewhere between the two. To the extent that our vision is blurred, we default away from harmony.

Simple men ask simple questions to which there are simply no simple answers. Man's destiny seems to be to crave for harmonic accord, whether found in nature, art or some other form invented by man as a precondition to living in tune with his surroundings. The world of the intellect must necessarily gravitate towards synchronicity with the universe.

A desire for, or lack of resistance to, a state of accord is fundamental to achieving harmony; but the means that man employs, ostensibly to gain these ends, are frequently anything but tranquil.

138 Aggression: Uncivilized Behaviour

The legacy of history has been that man has emerged through strife by prevailing over adversity and adversaries. Early man had no concept of civilization and therefore no concept of right and wrong except as events favoured or indisposed his personal interests.

Aggression for gain and as a pre-emptive defence strategy has become etched into his genetic codes. It will take far too long, expose life on earth to far too much risk and require more than faith to acquire a temperate benevolent mentality through the slow processes of natural selection offered by evolution. In any event, the requisite environmental conditions necessary for such a change to occur do not exist.

Physical Action

The physics of force are the means by which man and beast act out or counteract aggression. Whether the force is applied to increasing the distance between it and an adversary, or to assailing it, is a mental determination. Adrenaline is the common name for epinephrine, an adrenergic hormone excreted by the adrenal gland that, as the term suggests, autonomically adds energy by raising the blood pressure to amplify the forces that are applied to actions under aggressive circumstances.

Mental Control

Whether conscious or subconscious, the trigger mechanism of aggression resides in the neural system. Sensation tells the host of impending danger and the system responds. The quickest way to respond is to set the mechanism on automatic mode. All subconscious animals rely upon the automatic mode for no better reason than that they have no capacity to reason. Only man can judge the imminence and probable consequences of danger and condition his response accordingly. Only man can elect not to respond to the threat of danger. His response, whether active or passive, reflects judgement of his best interests.

Moral Standing

An animal is merely aware of the presence of an adversary, attempts to maintain a safe distance between itself and its nemesis and, if unable to, engages it aggressively. It has no scruples regarding the engagement or the possible consequences. Man on the other hand, having made a decision to engage, has the benefit of memory and the doubt of subsequent consequences. He will remember his assailant and, if he survives, take his adversary's prior performance into account on the basis of his experience. Doubts may linger with regard to how he may fare in future confrontations with this adversary if he allows his adversary to survive. Revenge is a primitive response attitude that prompts man to incapacitate his foe, for in so doing he averts the future possibility that he will again be assailed by the same adversary and perhaps not fare so well at that time. Judgement time is upon him, his own judgement. To react or over-react, that is his question.

Civilized society affords man assistance in this regard. It offers man protection under the law whereby if he constrains his impulses towards excess, limiting his actions to those permissible under the law, then the law will act on his behalf to deter and if necessary punish acts of aggression against him. Thus, man is placed in the delicate position of having to decide whether to act against his adversary of his own initiative and without regard for the law, a course for which the law may take its own retribution, or whether he should rely upon the natural deterrent of his own capacity for self-defence at some future uncertain date, at which time his personal strength as a deterrent authority would be augmented by that of the law.

To the question how one should deal with bullies; one should not! They are to be isolated from mainstream life. Under extreme conditions we do so with individuals and we can do so with states. Sanction is as sanction does!

139 Reciprocity: This for That

Of all that we have referred to concerning the behaviour of nature in reconciling imbalance, that which stands out most forcefully is the surety with which force is opposed by counter-force. When we apply this observation to relations between men we recognize that the flow of certain forces serves to bond them, others are seemingly benign and have little or no effect, while still others antagonize them. What distinguishes between the differing modes, and how does man control his conduct in a way that serves his best interests?

Bottom Lines

In addressing such questions we need necessarily to establish what it is that constitutes best interest. If, for example, we are intemperate and demanding, we tend to expect good to flow to ourselves at all times, and bad to flow only to others. But since that attitude is unlikely to deliver the greatest benefits over time, we need to establish whether our essential interests are immediate or longer in term; not to say that we do not entertain both, but rather we need to ask which frame of reference should predominate under any particular circumstance. This is the operative question before we ask any others.

The Wild World

In the wild world the present tense is primary for animals. If an animal does not act in support of its immediate interest, it may never have another.

Like the children's game of choice by which paper wraps stone, stone sharpens scissors and scissors cut paper (earlier referred to in discussing perception), the animal world is full of predators, each of which preys and is preyed upon. It doesn't matter where a particular animal stands in the predatory circle, it has no choice of weapons; and conscious choice of a strategy does not give the child any greater advantage in its game of choice in the absence of knowledge of the strategy of its adversaries. The wild world is a lonely place.

Animal Co-Ops

But some animals have evolved mutual aid strategies either to help their own kind or by forming unlikely tacit agreements to tolerate or even serve other species in return for services rendered; a practice called symbiosis. While this arrangement works to their mutual advantage, it is coincidental rather than self-chosen.

The Willed World

The question that now comes to mind is whether man, through the exercise of his intellect and will, can improve upon animal prospects of beneficial relations with others, and if so, how?

Inherent in all strategic planning is acknowledgement of the cause and effect relation. Implicit in this acknowledgement is the recognition that the correct response, the most beneficial rather than moral, depends upon precedence. One does not respond in the same way to two parties if one is hostile and the other friendly. Ideally, one would hope to find a strategy that was fail-safe, but to be so for all parties would necessitate parity, or at least relative satisfaction proportionate to the interests and expectations of each party.

Games People Play

To simplify the real-life impossibility of playing every option out on the ground, we model alternative plays as mathematical simulation games. The generic game that addresses this kind of problem, 'Prisoner's Dilemma', purports to suggest what the consequences will likely be if parties cooperate, mutually deceive or act with contrary motives.

Good Meets Good:

Where two well-intentioned parties agree to cooperate, in principle they share the benefits according to some amicable arrangement.

Bad Meets Bad:

Where two parties are intent upon deceit, they undermine each other's efforts and their combined gains are less than had they cooperated.

Good Meets Bad:

However, when 'Good' meets 'Bad', 'Bad' has few scruples and will steal all the benefits, if he can get away with them.

From the above scenarios the general rule suggests that whether 'Good' meets 'Good' or 'Good' meets 'Bad', he would be better rewarded were he to default and deceive the other in order that he may secure greater benefits than he otherwise would. This is the law of the Wild West, and of the big city!

The above plays are predicated upon two presumptions: that only two parties are involved in each interface, and that they do not expect to meet again. Under such conditions, doing wrong (to others) is the right strategy for securing greater rewards. The hinge of fate turns upon the contingent 'if' of getting away without being challenged.

If, however, more than two parties play, as in a Western posse or other form of alliance, or if the community is small enough that there is a high probability of reacquaintance of the parties, then interesting things start to happen that favour the 'good'.

Good Meets Good – Again:

Mutual confidence born of their first meeting further engenders trust. Trust begets continued cooperation, perhaps on a larger scale. The products: greater benefits!

Bad Meets Bad – Again:

Bad things happen, to either or!

Good Meets Bad – Again:

'Good' turns bad and bad things happen to 'Bad'. Revenge is sweet!

We begin to see a pattern whereby doing unto others as they do unto you begins to pay off. The incentive to deceive reduces proportionate to the probability of being deceived, until cooperation is discovered to be the best prescription for the most beneficial human relations. No sweat, no tears. Cheers! The process leads us from temptation and delivers us from evil.

The way we verify the correctness of this strategy is to observe what goes on in a human emergency. Absent incentives to deceive, mutual cooperation is not only the best but often the only recourse to salvation. In an emergency there are no temptations and no defections, only risks and potential rewards. The unspoken thought in any human emergency is 'there, but for the grace of God (or man), go I'. A state of collective operation has been co-opted by circumstances.

Correction

Haven't we forgotten something? Whatever happened to 'Good' after he turned bad to avenge 'Bad'? Do they continue to wreak vengeance upon each other? There must be an escape route and indeed there is. Under such circumstances, the option open to 'Good' is forgiveness. If upon reacquaintance with 'Bad', 'Good' offers 'Bad' an amnesty for all his prior bad acts in return for good behaviour, what are the chances of a turn for the good? Pretty good! Otherwise it is only a matter of time before 'Bad' meets bad luck.

But granting an amnesty to a deadbeat isn't going to help him get ahead. He is going to fall back upon his old tried and trusted ways of deceit. So in addition to the amnesty, 'Good' is going to reach down into his pocket and grant 'Bad' a token free meal, and some good advice: 'Make good or get out of town – or else!'

The ultimate power of reciprocity is its power to make good and correct bad through mutual exchange. That's good! The ideal parameters for this to happen are social stability and continuing opportunities for reacquaintance.

If nature was a little remiss in any way, perhaps it was that she did not provide for automatic and proportionate reciprocity for those who wilfully acted to damage others, including heads of state and military personnel. Apparently this responsibility has defaulted to mankind to provide the last piece of the relativity puzzle: the design and manufacture of justice!

140 Peace: The Linchpin of Civilization

When the hunters and gatherers settled to work the land they were spared the inevitability of encountering competitive hunters and gathers as they moved from place to place. Henceforth, if they stood their ground, competition would have to come to them. All they needed was the ability to defend themselves successfully against such eventualities. As any warrior will confirm, it is generally more economical in resources and lives to defend than it is to attack, all other things being equal. They had levered themselves into a position of advantage. The extent of their advantage could now be applied to building and manning defences, to herding livestock and to growing and harvesting crops.

Piece of Time

Relieved from the necessity to hunt, man was freed to apply his time otherwise. The time so gained could now be invested to advantage by designing tools of advantage: implements for tilling the land, vessels for storing the grain, devices for lifting and moving loads and defensive devices. With each advance in efficiency he gained additional time and security. Peace was stitched together piece by piece with time. A stitch in time saved nine.

Pace of Time

While the processes of improvement were advancing, so they were for others, the effect of which was to offset earlier gains. Adversaries learned of each other's gains. One cannot be unimpressed by the first sight of a wheel; learn fast to copy, apply and improve upon the lessons learned. The overriding lesson learned through such exchanges was the necessity to constantly change, with ever-increasing frequency. The accelerator was not designed; it evolved of its own momentous efficacy.

Peace Incentives

In order that peace could take root under such circumstances, one of two conditions needed to be present: either that adversaries should be of approximately the same strength, from which they would understand that their best interests did not reside in the war option, or that a state existed whereby one party was so dominant that no other could successfully challenge it. We have seen both these models in history but, as we have also seen, they were flawed. The first model depends upon mutual knowledge of the adversary's capabilities. This, of course, was not evident and in its absence the emotions lead men into battle where wise men should have feared to tread.

The second model relied upon restraint on the part of the stronger party from attacking its weaker neighbours. As we have already observed, warfare had become inculcated in the minds of emerging conscious man as a justifiable means to gain the necessities of life, and anything else of presumed value. The power of the powerful to prevail still prevails; the mechanisms of control are at hand, but the exercise of restraint on moral grounds is still wanting. The meek are not secure in the hands of the strong. Peace depends ultimately upon good will.

Clippety Clop, Bang Bang

If one encounters the assertion that 'might is right', one cannot but wonder why they are not spelled the same. More seriously, all Western movies depict the good guys as the better shots and the winners within some recognizable frame of reference. What is never revealed is why! Clearly, on balance, the ability of good and bad men alike to injure or kill with the gun is equal. In that context might is equated with maybe: maybe dead, maybe not. Right? Concealed in every Western is the social imperative for the good, the law, justice and public interest to be more than a match for its adversaries.

Peace Initiatives

What is missing is a clear picture of the relation between war and peace. Today we cannot afford to take any view of war but a world-view. It is the world that suffers the consequences of war, not just the combatants. To this end it is not sufficient for all nations or states to be as strong or stronger than their neighbours to deter war. That is not a practical solution. What is implied in order to maintain world order is that all should contribute to the security of all in proportion to their abilities and means. The aggregate abilities and means thus become the means by which to deter and overcome all acts of overt aggression between any state and any other state. In order to imagine the vehicle by which such controls may be exercised we tend, reluctantly, to recognize the necessity for a global strategy and strategic forces to stand in readiness to act in the collective interests of peace and at the expense of all – against miscreant parties. One of the costs, and arguably the hardest to bear, is acceptance of the decision to relinquish or dilute national or state sovereignty; the ultimate option being the release of claims to rights of self-defence by whatever means are deemed necessary. That is, in fact, not the appropriate argument or decision that would be required of subscribing members; they are only required to relinquish all initiatives to attack, except under the auspices of the global strategic force.

Any dissident states declining to subscribe and participate will become natural targets for other dissident states since the event of an attack would not automatically trigger the automatic response of the collective strategic forces. As more states join the alliance, the probability of non-member states coming under attack will increase until just one non-member state remains; at peace or frustrated by the absence of any other free states to attack. Economic sanctions and self-interest will eventually bring surviving renegade states into the fold.

The forces of international retribution should be so strong that no party can break the peace with impunity or survive without sustaining devastating consequences. One day, hopefully, war will no longer be a viable option because it will offer no viable opportunities. We will still, however, need to maintain vigilance and command over the forces of dementia.

What on Earth?

To abstract from the movies once again, this topic conjures up recollection of *The Day the Earth Stood Still* in which Michael Rennie plays an alien visitor to earth. The alien is accompanied by a robot that is programmed to automatically attack and destroy all weapons deployed aggressively towards alien citizens. The robot is otherwise unable to think or act beyond its programmed functions. Such robots police other planets. The concept of an automatic anti-weapon in an age when automatic weapons are so commonplace and so easily acquired has compelling attraction!

Imagine a world policed by a force that had no subjective or political allegiances and but a single mission: to overcome aggression wherever it arose. Such a force exists and is the ultimate weapon of choice by all those who aspire to higher states of civilization. It is called love.

Sooner, rather than later, the world must face up to the necessity of putting all its members on notice that an act of aggression against any member shall be interpreted as an act of aggression against all members. At such a time the world will act in unison to neutralize aggressors.

War is a frustrated search for peace; peace the denial of war or any lesser scale of conflict from revolution to rebellion to interpersonal expressions of aggression.

Peace on Earth

There is something very tranquil about being alone in a small boat on a large body of water, being wafted along by the vagaries of the wind and tide to a lap-lapping tune on the hull. It is a time when the mind is as free as it may ever be to reflect upon events.

Perhaps the skyline is broken by a millennia-chiselled shoreline upon which solemn sentinels stand informally, trees tall and dignified, exposing the vicissitudes of their past, quivering their finials in the present, while resolutely facing an uncertain future. Their stories, could they be told, bind them by analogy to one's own passage through time.

The stream of current events transports you from here to there, bending your mind as to where. In due course the tide may gently return you to your approximate point of departure evidenced by the return to view of your tall wooden friends and their peaceful relative alignments, as if to proffer an opportunity for a new departure to a brave new world.

141 War: The Failure of Diplomacy

Albert Camus suggested that a rebellion is an unsuccessful revolution and that a revolution is a successful rebellion. In either event they are states of disagreement with the state that have passed beyond the capacity of politics to resolve.

Rebellion, revolution and war are the contingencies to which we have earlier referred and politics gravitates when dissidence descends to violence. Camus notwithstanding, rebellion, revolution and war are descriptors of relative magnitudes of conflict in common understanding, regardless of any nuances that we may wish to embellish their distinctions with.

Causes

Disparity and despair are the root causes of war. Disparity in wealth and opportunities between peoples are the underlying seeds of discontent. On such occasions that we find ourselves in the same harsh predicament, as when natural disasters arise, we share our common humanity by helping each other to the greatest extent feasible. It is only when there is conspicuous disparity between those that have and those that need that discontent stirs the emotions to covet that which is there but not theirs.

Trigger Happenness

The predicament that man has always found it necessary to address is the question of how he can survive and thrive in an environment that is also inhabited by men of ill will and hostile intentions. Traditionally resolution has been afforded through force of arms and containment. The price was always very high but never so high that the survival of mankind was threatened.

Excrement happens! There are certain outcomes that flow as consequences of certain actions that are not intended, nor are they viewed objectively as attractive or desirable. Such events occur as auto-responses to the crossing of transparent thresholds of human tolerance, as if crossing the fulcrum point of a seesaw to transfer the balance of power from one authority to another. At these thresholds in human affairs, the emotions take charge of reason to create instability and unpredictable behaviour. The maintenance of peace no longer rests upon cool deliberation and self-interest in security, but upon whether or not certain auto-responses kick in to take charge; and whether there is enough kinetic energy pent up behind the discharge to frustrate or countermand established order. The emotions are the trigger mechanisms that initiate the change of mind and minds from relative happiness to a state of happenness, as of war.

Consequences

The execution of war returns governance to the controlling authorities. Control is the exercise of power by design. Design is the product of planning. Planning is a process of reason. While the emotions still feed the motivations, it is cool calculation that supplies the means to curing and securing an end to social insanity. The tempest has been unleashed upon man by man as a consequence of his not having provided the requisite measures necessary to maintain peace in the first instance, and of loss of control of the emotions in the second. Social cohesion only binds peoples together when the glue is spread generously. The effect of armed conflict is to diminish the capacities of adversary forces to maintain armed opposition. The parties are, however, never returned to the same state of relations that existed before the event.

People who demonstrate war-like behaviour are the same who should justly suffer its consequences.

Costs

What they have gained, regardless of the outcome, is first-hand acquaintance with the wretchedness of war as a means to gaining ends. Payments are made by the victor and vanquished alike, in spent lives, resources and emotions. No party wins a modern war; each loses more or less relative to the other. Things are the way they are for very good reason, even when our reasons are not very good. We arm ourselves to the teeth to blow away our teeth!

Benefits

For better or worse we seem to gain better experience from worse experience. We learn more valuable lessons from the negative than from the positive. What not to do is the best worst lesson! The whatnot, meaning the unstated, is not 'what' but 'what not' to do. What 'not' to do is the best question. When one has exhausted all the 'whatnots', one has the best 'what' at hand and in mind.

To War or Not to War?

The determination as to whether or not to engage in warfare is, for most, traumatic. The simplest decision arises in response to an unprovoked attack. Like an individual, a sovereign state reserves for itself a sovereign right to defend its physical integrity. With regard to mental integrity, ultimately war is a state of mind. How one casts the mind from a peaceful state to a state of hate depends in large measure upon the degree of threat perceived. Therein lies the difficulty. Absolute hostility translates to absolute reciprocity. But what is the appropriate response to an increasing potential for aggressive behaviour in the absence of any overtly aggressive action? Logically, degrees of threat warrant comparable degrees of preparation for defence and counter-attack for the purpose of deterring or disabling a party's capacity to successfully take aggressive action. The logic of such situations is clear: active force invites equal or greater force in reaction. So far the extent of necessary preparation appears clear.

There is, however, another form that hostility may take: the development of means of aggression that, if used, would render any reciprocal reaction moot or impossible due to the extent that an initial attack would incapacitate a targeted community. The development of such capacity appears to give the would-be aggressor a double advantage: first by affording a means by which it can coerce its adversaries and, second, should that not achieve the intended results, the implementation of such means while minimizing negative effects upon the attacking state.

The question thus arises as to what options best befit the interests of peaceful states when confronted by threats to national or international security. Clearly the tacit acceptance of such a state of pending calamity is not acceptable. We have earlier treated the subject of reciprocity and ultimata: this-for-that strategies of exchange or else! We have also spoken of the subject of trust; that it should be extended proportionate to one's previously experienced relations.

The prospect of building a system of deterrence may be ineffectual if a potential aggressor cannot be relied upon to act rationally. Therefore, with much regret because it augurs a fundamentally new interpretation of the conduct of ethics, one prospective response suggests itself: the pre-emption option!

Like the strategy of the would-be aggressor, pre-emption can take two forms: first coertion whereby peaceful states issue inducements to a would-be aggressor to cease and desist from engaging in activities that can be misapplied to conduct calamitous wars. The 'or-else' option would be, for such states that deem themselves to be exposed to unacceptable risk, to stage a pre-emptive attack upon the would-be aggressor before it gains a capacity to so endanger international peace and security.

This last act is not as fundamentally novel as it may appear to be. War has always been waged by parties that believe that they hold a decisive advantage over their adversaries. Yet such circumstances as may render a target state totally disabled cannot be interpreted as an act of war in the conventional sense since there is no residual capacity to react! There is, in fact, no war as we would traditionally define it! We reserve such adjectives as cataclysmic and catastrophic to describe the effects of an over-preponderance of force discharged by one party upon another.

In the event of a decision to incapacitate a would-be aggressor, the hinge of fate therefore rests upon success in the application of said inducements or a pre-emptive attack, both of which presuppose an ability to assemble a system of overwhelmingly superior controls over the rogue state. To that end, the greater the number of free states that act in unison, the greater their authority and prospects of a successful and speedy outcome. The unity implicit in internationalism thereby becomes the ultimate weapon for peace.

To this general strategy for peace we should add an important caveat: that the combined moral, legal and military authority of the international side of the equation should always significantly exceed that of any single state, thus denying a single state the temptation to become an aggressor.

An anomalous cloud hangs over the issue as to what circumstances confer moral and legal justification in the taking of pre-emptive action by one state against another. The answer with regard to taking such action appears to hinge on a judgement that not to do so would, in all probability, result in calamitous consequences. Such a decision, if and when taken, would be indistinguishable from an act of patent aggression when viewed by third parties unless supported by verifiable evidence confirming that the rogue state is an imminent threat to others.

To War

Confucius was convinced that war was evil. If and when it must be undertaken he maintained that it should be conducted vigorously by persons who have knowledge of why they are fighting, believe that they do so for a just cause and are convinced of the necessity for success.

The humanities provide the grounds for just causes while science, mathematics and industry provide the means to just ends. In the absence of a morally justifiable 'why', no 'how' is justifiable.

The conduct of war is an ugly scene from any perspective. One thousand years ago mankind killed his kind one at a time, and most of the time eyeball to eyeball! Conventional wars have brought into confrontation sides that are approximately equal in their inventories of armaments: sword to sword, tank to tank and battleship to battleship; a fact that goes a long way to explaining why conventional wars have generally been won by the party with a preponderance of numbers.

The obvious explanation may be found by referring to principles of mechanics by which a greater force overcomes a lesser force. In attendance with and running parallel to these general laws is a rule that demonstrates a form of social justice that we have slowly come to recognize, adopt, cultivate and name democracy.

Dogma denies doubt. In the absence of dialogue dogmatic regimes persecute the dissident and engage rival ideologies in war as the only option open to resolving their differences. War in an era of compounding technological prowess only accelerates doomsday for life on earth.

Today technology has rent great disparities between warring parties whereby individual targets can be reduced with surgical precision or broad areas laid waste by an adversary that is in no danger and thousands of miles away from the event. Thus, troublesome questions arise with respect to how majorities can maintain superiority on the one hand, how the greater good can best be served and whether these interests may, upon occasion, be at variance, bearing in mind that majority rule and justice do not invariably coincide.

Not to War

A model code of conduct requires to be engineered by a critical mass of mankind that stipulates that those that threaten the peaceful coexistence of others shall have their own peace so threatened. This does not sound fundamentally different from the standards that we practice today except for the distinction that majority rule necessarily demands a means of ultimate control where the interests of humanity are threatened. By inference we draw the necessary conclusion that no single faction should be free to unilaterally pursue policies that threaten the survival of others. To that end, if mankind were to establish and maintain a collective position of military superiority and readiness at all times, the issuance of a warning, followed by an ultimatum as necessary and required by a pre-established protocol would be all that would be required in most instances to keep the peace. In the event of a renegade state initiating conflict, the counter-action, use of preponderant force, should be automatic and decisive.

What appears to be necessary for the security of peace is a contingent threat of reciprocal destruction. Fear trumps arrogance and aggression.

Wars are instigated as means by which to gain ends that are disproportionate to deserts. In a war between fanatics and moderates it is easy to assume that the fanatics, being blindly committed to their self-righteous cause, would win. However, history belies such an assumption. What war was ever won by the impassioned, self-inflicted blind? They can hardly be expected to act rationally or elicit the sympathies of otherwise disinterested third parties.

A great hunger for social and personal security follows wars. Men stand in defiance of aggression to reassert their prior peaceful states. In an ideal world, peace is analogous to a state wherein prerogatives to access the blessings of life would be approximately uniform. No critical disparities, no wars! Wars of aggression undermine the realization of such high ideals.

142 Arms: A Farewell to Welfare

A few words are in order with regard to claims of rights to bear arms. As we look backward in time we can readily identify with the needs of individuals, families and groups to protect their lives, properties and interests from acts of violence. We can also recognize the advent of collective security arrangements whereby the defence and protection of community interests became the responsibility of qualified militia trained and practised in that undertaking, later supplemented by local police drawn from the community.

Given the extent to which these institutions have become inculcated into the fabric of modern societies, the question sooner or later naturally arises: what purposes are served by continuance of the prior system that sanctioned the ownership and use of arms by individual citizens?

The processes of civilization move mankind further from the contingent necessity for self-defence. Indeed, the extent to which man continues to feel the need to retain his option to carry arms is a clear signal of how far his society has progressed, and has to proceed in the process of becoming truly civilized. We reflect upon a similar truism whereby 'the greater the number of laws on the books, the greater the distance that society stands from being truly free'.

Welfare

The aspirations of mankind to fare well are fairly well understood. The welfare of people and their societies is fundamental. To the extent that we deny others the fulfilment of their inherently natural aspirations, we deny our access to higher destinies. What we fear in others, others fear in us. How those fears play out, particularly with respect to our actions in relation to others, is what we invite for ourselves. Welfare is not a government programme; it is a programme of government whereby the wellbeing of all beings is a paramount determinant of the actions of governors and the governed alike.

Farewell

To the extent that we feel the need to point guns at each other to coerce, or worse, we distance ourselves from a state of wellbeing. Goodbye future, until sometime by and by when we will be ready to accept the greater good of that future. Meanwhile, as erstwhile, we shall hold arms in reserve for contingencies for a long while!

The two postulates upon which the existence, retention and use of arms rest are the conquest of whatever is and the defence of it. Until we rise to a state where we can cast off these defects in our natures, we shall no doubt retain the right to bear arms. Trigger-happiness is a consequence of profound psychological unhappiness.

The reason that guns command the attention of so many Americans is that American society is inherently very insecure. Guns represent a means by which to offset instability, which outsiders correctly perceive exacerbates it. 'If you want to understand a relationship stand apart from it.'

Article II of the Amendments to the US Constitution stipulates that 'the right of the people to keep and bear arms, shall not be infringed'. This right is predicated upon the circumscription of 'a well regulated militia, being necessary to the security of a free state'. No other constitutional rights pertain to the holding and bearing of arms, nor are the types of such arms specifically unlimited. This being so, Congress is within its rights to limit the holding or bearing of arms of any type, and their use for any purpose other than in support of a well-regulated militia, their use against a militia not so well regulated or for self-defence.

In time the interests of those who wish to be protected against violent crime will overbear the interests of those who want to sell and use guns, and of those reluctant to entertain change because uncertainty is more disconcerting than the probability that guns will be used against the innocent. What is more uncertain and infinitely more tragic is the number of human sacrifices that will seemingly need to occur before a majority of citizens takes up the call against arms.

143 Success: The Sunset of a Goal

Success

How often we hear the word success as a declaration or acknowledgement of achievement, and yet how seldom we hear a clear description of intent before the success is achieved. Granted, intentions are often presumed, as if all accomplishments that end with a measure of pleasure had started with a dimension of intention. But is this so and, particularly, is this so in matters of common interest? We are all economic animals that operate in economic environments, yet very few of us set out to make money. 'Oh yeah' you may say in disbelief!

But do you know anybody who actually manufactures currency by hand? We go to work to do other things by which we hope to gain necessary but often secondary by-products, such as money.

When someone acquires a conspicuous accumulation of wealth we are quick to proclaim his 'success'. Our culture equates success synonymously with the accrual of money. Yet how often do we enquire about the degree of satisfaction that has been gained from pursuing money as a primary occupation? How often do we understand whether someone has fulfilled a deeply felt and significant need? Do we adopt goals for the purpose of making us unhappy? Is the security of happiness not a better cultural analogue of success?

Those who boast of their success are generally referring to their financial success. It likely takes a third party to recognize and confirm achievements of other kinds. What one may view as a success does not necessarily correspond to the intentions or understanding of another who enjoys similar rewards. Thus, the question arises whether accomplishment can be defined as successful in the absence of intentions directed to such ends.

Success is the sunset of a goal
Without such definition leaves a hole
Integrity depends upon the wholes
As necessary attributes of souls

When the wholeness of the goal is only token
And commitment runs no deeper than words spoken
Then the sum of parts is greater than the whole, and
Failure is the sunset of a goal.

Failure

So we discover that the converse is also true. In the absence of a goal one cannot fail. This should indeed be reassuring for those who didn't manage to rise in their own self-esteem or in the esteem of others! Apparently if you don't set yourself up to win, then you don't lose. Extraordinary as it may seem there are many who are quite content to assume a passive role observing life. Who is to say that they do not choose to be observers, and having so observed life's rich variety, who is to say that they have not succeeded by some meaningful measure?

In Between

What do you call someone who is something between a winner and a loser? A human being being human perhaps? We are all winners in small ways, and all losers in more ways than we like to admit. The fact of survival is the most creditable of successes, particularly in a competitive world where 'fish eat smaller fishes eating smaller fishes eating' etc.

So there we have it again. Each life has but one ultimate failure, death; and all other accomplishments are, by comparison, relative successes as expressions of survival.

We live lives of serial decisions presumed to be in our best interests. For most there is no overreaching goal, there is just expedience, and it is invoked now, and now and now.

To Will Success

It is a truism that the higher the goal, the greater the challenge, and that the greater the challenge, the less likelihood there is of achieving success. This precept seems to account for the greater number of lesser goals and the relative paucity of virtuous achievements of a very high order. In this regard the exploration of territory believed to be inaccessible to the human intellect cannot be discounted as an unworthy goal merely on the grounds that it cannot reap tangible rewards.

Those are the caravelles of the Don Quixotes, Candides and Little Princes of this world who raise the spirits of the rest of mankind through revealing eternal verities. 'Heroic love is the property of those superior natures who are called insane not because they do not know, but because they over-know'; choice words from the ex-Dominican philosopher Giordano Bruno.

A caveat must therefore be introduced here that appears to contradict the inference that the pursuit of goals is virtuous. Wisdom cannot be gained by setting as one's goal the objective of finding wisdom. If we hold that wisdom commands the high ground of moral achievement, we must relegate the approbation of goals to the status of will whereby we commit ourselves to journeys, the destinations of which are often unclear. It is the act of cause, rising to a commitment that invites a corresponding effect.

Paths to successes are as diverse as the goals towards which they lead. To win in competition may require one to excel in the arts, academia or athletics; to be the cheapest in bidding for work; to overstate the truth in politics and the law; or to understate feelings in diplomacy and in the face of human tragedy.

What does all this tell us about goals and success if not to suggest that, as we reflect upon our successes and failures, it should not be to success that we give all the credit? In equal measure credit is due to the identification and maintenance of very high, even unattainable goals that, by definition, ensure that success cannot be realized in absolute terms.

Perhaps the ultimate success is to do something so well that nobody is quite sure what was done, where, why, when, how or by whom, as if the act precluded man. God knows!

144 Welfare: To Fare Well

If the first thing that comes to mind on reading the word 'welfare' is a government assistance programme for indigents and other unfortunates, you may need a little assistance.

Individual Welfare

To fare well is to enjoy good fortune. As to where good fortune comes from is a matter for some consideration. Intellect, physique, imagination, tenacity, shrewdness, indeed countless hereditary traits are gifts for which no credit is due those so favourably gifted other than the fact of their exercising them of their own initiative, which may also be genetically sponsored!

There is another faculty that we cannot attribute with confidence to heredity but which is characterized by the ability to find favourable things not sought. Perhaps this is simply a reflection of the experience of those who, by luck of circumstance, blunder into beneficial experiences but who would also necessarily have needed the disposition for being receptive to recognizing the merits inherent in such events. This quality, derived from the Sanskrit word *simhaladvipa*, meaning Dwelling-Place-of-Lions Island (modern Sri Lanka or Ceylon) and corrupted through Indian to the Arabian *Sarandib*, first found expression in English as 'serendipity', a word coined in the eighteenth century by Horace Walpole in his interpretation of the Persian fairy tale *The Three Princes of Serendip.*

There are sound grounds for crediting many of our good fortunes to others. It is to others that we owe our debts in return. It is not what you have that tells others who you are but rather what you do with what you have. That is a function of will and work.

Collective Welfare

Can we in good conscience sanction the rules of chance as the best, the most just basis for compiling the laws of civilization? Should we thus interpret the rites of birth as birthrights?

Ought we therefore embrace inherent privilege as the natural and proper qualification for leadership regardless of capacity? History has painfully answered these questions in the negative.

The state of faring well is assessed by comparison with the circumstances of others. It boils down to the question of who is faring well.

While the United States may lead other countries in its economic and technological developments, a society that has more than 40 million citizens without access to medical services (more than the populations of Denmark, Finland, Norway, Sweden and the Netherlands combined, all of whom enjoy such access) cannot be construed to be faring well! One cannot but wonder what the purposes of national economies and technological wizardry are, and what kinds of people they are that don't care for their own kind when the means to do so are so readily at hand!

Correlating material wellbeing with spiritual wellbeing often suggests that a surfeit of one signals a deficit of the other. This interpretation does not hold of necessity but rather tends to reinforce predispositions towards the defence of prejudice.

The authority behind the 1960s maxim 'question authority' arose from experience: the experience of not faring well or being well served by trusting God, experts, institutions and government. From appearances the proponents represented left-of-centre politics but the irony of their claims was that they were to confirm just how 'right' they were in verity, virtue and aspiration for the individual, trusting their own judgements rather than the homogenized hearsay and blind dictates of the doctrinaire.

As to the government welfare to which we initially referred, it is part of the obligation of the fortunate under the social contract, a privilege of the privileged one might say, to provide support as may be necessary to assure the security of the less privileged, thereby underwriting their own.

145 Authority: The Power to Control Events

The most critical aspect of addressing the subject of authority is the determination of a system of classification appropriate to the needs of the subject under consideration. Whether one is concerned with the form of government (or state having such form), with the mode by which government is constituted (distribution of sovereign powers), with social or political classes, theories of organization or with some other basis of comparison of power apportionment, there are always alternative manifestations of authority that challenge any generalized system.

What distinguishes authorities in general terms are numbers – the one, the few or the many who control events – and whether or not they serve the general good.

Plato Said

Plato held that virtue was the correct standard by which to classify governments. By this standard he ranged the alternatives (from worst to best) as tyranny (of the most corrupt), oligarchy (a few corrupt), democracy (many without virtue), timocracy (many with virtue, meaning value, meaning honour or wealth), aristocracy (a few with virtue) and autocracy (the most virtuous).

The problem with this classification was two-fold: first, that, because an autocracy (rule of one) was monopolistic, it tended towards corruption of the interests of others, the distinction between the most virtuous and the most corrupt being difficult to draw for someone who was aggrieved by the actions of the leader; and, second, that virtue was predicated on the basis of individual behaviour rather than, as we prefer today, upon a system of just sharing of powers based upon varied aspirations and mixed virtues which system itself is held to be inherently more virtuous. These objections could to a large degree be overcome if the verb '(to) rule' was conditioned by the noun '(a) rule', meaning (a) law.

The ideal seemed to lie in a delicate mix of the virtues of each system that would maximize the benefits while minimizing the risks. Plato propounded that prescription in *The Republic*, a form of government that has served as a model of social organization for nearly 2,400 years, further qualified by his later work *The Laws*.

The Virtue of Virtue

Insofar as we have recognized the critical importance to civilization of security and advancement, these two criteria may be broadly identified with the good. The process by which we advance virtuous causes is to secure the instruments of progress. Thus, if we recognize the potential value of the sage to society, we protect this asset from debasement, affording him every opportunity to contribute to society in the manner best befitting his talents within the bounds permissible by law. The same is so with creative minds generally; artists, poets, inventors, scientists and members of the learned professions, all of whom function optimally as individual engines of change for the good. So the issue of need to protect and nurture virtue is not in question but rather a goal. The real question is how. How can we facilitate the virtuous authorities of one while denying the emergence of tyranny in another?

The virtues of the few may be similarly demonstrated by showing how the integrity of small groups working in unison to promote public health and safety, education, industry, etc. are essential to the wellbeing of society; while in matters of interest to all people, all people can and should participate by exercising their authority on an equal basis – as is the case in voting in a democracy.

The Virtue of Numbers

Numbers command. Ultimately the number of people committed to a cause compared with the number committed against it will resolve in favour of the greater number, all other considerations being equal. This principle is the greatest security against tyranny, but only if protected by laws and the power to secure laws. Security protects security protecting security.

Numbers have a diluent effect upon the beneficial contributions of individuals and small groups. With this in mind it is important that each unit of scale be granted the greatest opportunities to flourish, subject only to thresholds of acceptable behaviour governing the interests of society in curbing vice and corruption. It is noteworthy that Plato's advocacy of a classification of authority based upon virtue not only places the rule of the many far from virtueless tyranny, but equidistant from exemplary virtue! We are to be persuaded that the understanding of the average man should constitute a higher authority than the perspicacity of an exceptional individual!

Autocracy

The state of wild animals and of wild men is that of not conforming to the dictates of external authority. They live in accordance with the law of the survival of the fittest and vie and die on that basis. Jungle 'lawfare' provides for all to be governors of their own actions.

As man congregated to forage and pillage, the strong, led by the strongest (and most ruthless), commanded the weak. This was government of the many by the one. It worked because the strong were constantly challenged to maintain their positions through demonstrating their superior abilities to prevail. Those that rose to challenge authority were governed by the same tests. Those that did not prevail did not survive to plead their cases. The concept 'might is right' arose from confusion between the concepts of fact as reality and fact as verity. This confusion was fuelled by the experience that some leaders were virtuous while others were tyrants, the very best and worst forms of authority.

As the many became too many, settlement and emigration from settlements led to the founding of new communities. The one-to-one relation of settlements and their leaders led to competition for resources and the mindless warring of which we have spoken.

Theocracy

At times of overwhelming stress, mortal combat being such a time, men 'listened' to their inner 'voices', believed them to be 'heard' and attributed them to the external authority of gods. They organized society on the basis of what they interpreted and agreed the gods had 'said' and 'intended', and acted upon those beliefs because they had no better rationale for organization.

Communions of like-thinkers formed the organized religions of latter days that have governed and influenced societies' mores and largesse, more or less, to the present day. The distinct character of theocracies has been the presumption of a monopoly of virtue. With this banner flying on their staffs, surely they were invincible! And so they charged into battle. Crusades in the name of religious authorities have been shown to be as despotic as any autocracy, their authority eventually weakening along the ever-extended boundaries of their imperial holdings.

The term 'state' came into use to describe the secular equivalence of ecclesiastical authorities following the sixteenth-century Reformation movement, the breach with Rome and subsequent movements to separate the governance of secular from clerical affairs. Thus, the term 'state' referred to a 'state' of not being governed by the church. Of profound significance, the rigorous religious, through their professions and actions, implanted the seeds of a dissolution of political boundaries about which we will hear a lot more in the future. The continent of South America is still a de facto Roman (Catholic) colony. 'The king is dead. Long live the king!'

Aristocracy

The distinction between theocracy and aristocracy is theological. In all other respects their interests coincide. They are both plutocratic squirarchies, i.e. wealth and land acquisitive. As to whether aristocracy is virtuous or merely so-styled (from the Greek *aristo* meaning the best) to distinguish it from the oligarchy would depend upon what they do, understanding that much of what man does much of the time, boozing, schmoozing and snoozing, is not virtue-significant!

While a sovereign may well wield absolute power, his maintenance of absolute power depends upon an oligarchy or aristocracy. Whether he is a tyrant, a sage or something in between makes of his system of government a highly volatile mix of vice and virtue – a poor bet and a poor source of security for his subjects in the long haul.

Totalitarianism

Dictatorships that ostensibly claim power in the name of the people have the distinction of holding that their ends justify their means. The means by which they come into power are invariably corrupt while clouded by pretensions of not being so.

The significance of totalitarianism attaches to what it is capable of doing. The powers of the state under the single-minded direction of fanatic leaders enable acuity of focus upon goals that is not possible in more egalitarian societies. In recent history the Russian Soviet, German National Socialist and Chinese Communist revolutions have advanced the statures of their respective people militarily and technologically at rates that are not possible within the democracies simply because totalitarian administration is not compromised by the necessity to address issues of virtue. Their peoples have lent support to such regimes in the mistaken belief that they had no choice and that these systems offered them the only means of advancing their collective welfare.

Democracy

Democracy as defined by Plato and Aristotle embodied no presumptions of virtue. It was a practical expedient that arose as an optional form of government simply because it was possible. It was possible to convene occasional meetings of citizens (not including women, children or slaves) for the purposes of exchanging ideas and voting when the political unit was in the order of thousands (not tens of thousands) of eligible citizens. Direct democracy was well suited to the scale of the *polis* or city-state. The democratic experiment of Athens was, however, shortlived and democracy did not arise as a cogent mode of social organization for another 2,000 years.

When democracy did re-emerge it was in seventeenth-century England, arising on two fronts in the Age of Enlightenment. The first was through written word, notably by John Locke in advancing his theories on the need for the separation of legislative and executive powers of government, the need to protect the voluntary status of religious beliefs of minorities and the necessities for guaranteeing 'natural' rights to life, liberty and property. He was later to write the Constitution of the Colony of Carolina. The second front was the slow erosion of the absolute power of the monarch being advanced by the emergence of concepts of 'natural law' (meaning reason), at first to bring logic and legislative law more closely into correspondence, and secondarily to introduce what we would today call greater fiscal responsibility to the processes of collecting and distributing state revenues. In small increments over the next 300 years the people claimed and acquired title to sovereignty from their sovereign. The monarch was to become, de facto, a non-executive president.

While it was Locke that articulately refuted the principle of the divine right of kings, we should recognize that experience, conspicuously scientific experience, is refuting current and popular states of knowledge, and therefore values, on an ongoing basis. Today the West enjoys an almost hegemonic application of the principle of democratic rule. To appreciate the fitness of social mores to particular purposes, we need to exaggerate their effects. A case in point would be the administration of war, which if conducted on the basis of democratic principles would surely deliver defeat more often than victory.

The 'flavours' of democracy that have emerged to date are direct democracy, whereby citizens participate in all decisions; representative democracy, by which they elect proxies to represent them; constitutional democracy, wherein certain rights are guaranteed and protected; and the twentieth-century brand, social (or) economic democracy, which finds expression in claims of rights to collective welfare. In any event, democracy embraces a state of interdependence. The concept of democracy is not, however, as simple to grasp as those who enjoy its benefits may assume. In the autumn of 2000 a Lebanese newspaper published the following apologia when challenged to explain an error in its reporting: 'We are used to a deep-rooted Arab tradition of democracy where results are first declared, then elections are conducted and votes brought in to affirm it.' Alice would have felt on familiar ground if confronted by such *Looking Glass* logic.

Socialism

If by socialism we intend the governance of the people by the people for the people, socialism as an ideal is irreproachable. It is only when we attempt to put the words to the music of practical application that we reproach ourselves for ever having entertained such a good idea. It is not that we are wrong in entertaining high goals, we are simply not up to the challenge of making such a system work well at this particular stage in our social development.

Twentieth-century experiments with communism have usurped the 'words' of socialism as pretexts and professions of ends to make 'music' for the ears of the 'few' at great cost to the 'many'.

Like Athenian democracy, socialism has worked best for small, homogenic and relatively stable populations. Like Greek democracy, it may have to wait a while before rekindling its fires in the breasts of mankind. Perhaps we need to beat up on each other a little more first, or to encounter adversities that are so overwhelming that we have no option but to collaborate.

Forms of Government

The principal function of government is to address the subject of security. As to whether security is defined narrowly as the protection of the leadership or the state from aggression, or more broadly as the welfare of the populace, is the subject of ongoing factional debate. The competitive forces that define the nature of a government are functions of consent and constraint.

How governments are constituted, whose values they represent, how social order is maintained and what mechanisms are adopted for their effective operation vary in mix between the extremities of absolute control and absolute anarchy, the latter being a state of absence of effective authority and therefore of capacity to govern.

The parliamentary system is an outgrowth of council (parley) to, later with, and subsequently without English kings. This system, much emulated throughout the world, gradually evolved in the direction of transference of power from the 'one' to the 'few' to the 'many', becoming, as we have noted, the vehicle for the rebirth of democracy.

The federal republic form of organization patterns itself around the federation of (e)states into larger units of government while retaining limited powers of self-government in each, the term republic referring to a system having a chief of state who is not a monarch.

Under democratic forms of government (i.e. parliamentary democracies and democratic republics) transactions in 'goods' (virtues) occurs multi-laterally. Thus, the next order of business was to forge a mechanism to secure this social advantage for posterity.

Constitutions

The laws of a simple democracy, by popular definition, disadvantage minorities. The processes by which authority is constituted and reconstituted are therefore critical to the form that authority takes. The distinctions between constitutional means turn around the central necessity for controls. Thus, the intrinsic structure of a constitution determines to a large extent whether the authority system is virtuous or perverse.

Clearly an authority of one is centralized. There is no need for formality of organization. As individuals we seldom deliberate regarding what form our decisions may or may not take. Our individual constitutions are a function of internal values and subconscious or conscious dispositions.

Where the interests of many are at stake, the issues are more complex. First, there are the all-embracing interests in common, like security from external aggression and general welfare (however that may be defined); then the interests of lesser numbers who have agendas that may conflict with the vested interests of the larger group; and, lastly but no less important, there are the interests of each individual.

Some of these interests are so compelling as to merit general application without exception, as, for example, the interest in the protection of life, liberty and, to a lesser extent, property. Other interests merit discretionary powers that would be vested in certain individuals.

What therefore presents itself is a need to define such essential interests, to classify them with respect to the extent to which they are more or less generally applicable, and then to arrange them in a hierarchy whereby some are afforded the status of being inviolate while others are less rigidly defined and protected by degree.

The principle 'less is more' reasserts itself by which authority operates efficiently, always understanding that efficiency is not the end goal but rather one means to gain virtuous ends. In each decision context involving authority we should therefore ask ourselves: what is the most efficient increment of authority that can achieve the intended purpose without undermining the general principle of virtuous intention? It is not the intention per se that delivers the virtuous result but the act that is directed by such virtuous intentions. So the purpose of a constitution is to establish fundamental rules and principles that determine the powers, duties and rights of people.

Model Constitutions

We shall confine discussion here to constitutions that address the organization of states. We have outlined Plato's ideal constitution that was itself reconstituted by Aristotle as a model.

Since we have established that the purpose of a constitution is to forestall tyranny, and since we can readily identify certain cultures whose constitutions do not stand up to that test, we shall focus here on the constitutions of two modern democracies: one that evolved slowly and painfully over nearly 1,000 years, and another that was conceived for a particular purpose at a particular place and time in the light of the strengths and weaknesses of the former.

The formal beginnings of a constitutional stream of thought in England can be traced to the early thirteenth century. In the year 1215 a coterie of barons presented their grievances, demands and an ultimatum to King John in connection with extortionate dues, fines, imprisonment, pillage and land seizures by the crown for failure to pay. Under threat of war they demanded that the security of certain rights be protected under the law in perpetuity. At a meeting in Runnymede, King John acceded to their demands in return for their continued loyalty. The Magna Carta was signed by the parties and reissued six times before being confirmed by King Edward I in 1297 and read into the law of the land.

The British constitution of today is a complex of statutes, political party prerogatives, conventions, traditions and other crowning glories that, if written down, would make even a monarch blush – but its true measure is that it works as well as any other.

The circumstances that led to the rebellion of the American colonies against British rule more than 500 years after Magna Carta were not, therefore, novel but rather due to forgotten or disregarded lessons of history.

The Constitution of the United States of America was written by dissident Englishmen following presentations and rejections of their grievances and demands that culminated in the Declaration of Independence and the Revolutionary War to establish that state in fact.

The salient distinction between the British and US constitutions was the centralization of power in the constitutional monarchy compared with the decentralized format of a federal republic that established geographic subdivisions, guarantying a broad measure of jurisdictional autonomy to each. While all governments distribute powers in some format, the special case of the US was to decentralize power as a constitutional right, a reality that necessarily demanded a different form of government appropriate to the large scale of the territory.

The great virtue of constitutional democracy is that it places controls on abuse that are not simple to overturn. In so doing it protects so-called inalienable rights.

The Principle of Separation of Powers

We shall use the clearly differentiable mechanisms used to control abuse in the US as a model for the purposes of illustrating the intent and effect of the separation of powers doctrine. Reflecting the inherent stability of the triangular relation in geometry, the US government is divided into three quasi-autonomous branches: the legislature, the executive and the judiciary. The legislature or body of lawmakers has two conflicting principles to uphold: the advancement of democracy through proportional representation, and the protection of minorities. To this end the legislature is divided into two bodies; the House of Representatives that, as the name suggests, deliberates through elected representatives in the interests of majority rule, and the Senate, whose members are elected to represent each state in equal numbers (two per state) with equal authority.

The executive is charged with administering the day-to-day business of government. It operates on a hierarchical top-down principle. Elected 'somewhat' democratically, the president presides over a cabinet comprising heads of departments, and over specialized agencies. (We shall address the 'somewhat' somewhat later.)

The judiciary is entrusted with the direction of the administration of justice. Their domain carries down from the rendering of judgements in the courts, to the security of prisoners and the enforcement of laws through police action. This is the only branch of government that is not held fully accountable to the other branches or directly to the people except as they may jointly elect to amend the Constitution, an act that by design is very difficult to accomplish. We shall speak of the special character of this aberration in the exercise of checks and balances in government when we discuss justice in a later section.

What we call bureaucracy is the system of elected or appointed staff, including professional civil servants, that performs the work of each subdivision of government.

Actions of Last Resort

The importance of the natural interests, logical rights and practical needs of the people to override the legislative process in order to circumvent intransigence, complacency or corruption in government should not be underestimated or denied when the clear and decisive wish of the people favours a specific remedy. To this end forms of plebiscite or direct democracy, the recall (to remove elected officials), the initiative (to amend legislation by popular vote) and the referendum (to impose legislation by popular vote) have been instituted as intermediary relief valves to forestall what would otherwise be the last resort of peoples frustrated by their elected officials: their expression of displeasure in the forms of civil disobedience, rebellion or revolution.

Conversely governments, acting in the interests of the people, reserve the ultimate authority of rights to foreclose private properties or operations if the interests of the public clearly outweigh those of the private owner, the matter of just compensation varying by jurisdiction. Such interests are well exemplified by monopolistic practices that extort or otherwise detract from the interests of the public domain – hence the term 'eminent domain' used to describe such foreclosure rights.

The veto is another such case in point. The veto is an exercise of eminent authority over action that is deemed by one branch of government to trespass upon the preserves of another; or may be invoked by a top executive officer (as a president) to deny contravention of a constitution or the passage of legislation into law if it is held not to be in the interests of the nation or state. Whether a veto is absolute or may be overturned is generally a function of constitutional law.

Man is an overgrown child, and like any child he needs limits prescribed by society to contain his natural primitive tendencies towards self-indulgence at the expense of others. Perhaps Hobbes's political instincts (that antagonized proponents of the social contract and the monarchy of his day) have been vindicated by our concurrence and actions in insisting that some form of constraint mechanisms are necessary to curb excesses in all aspects of authoritarian human behaviour.

Internationalism

Where we have no local options, as in dire emergencies, internationalism (i.e. worldwide cooperation) takes command. What is of lesser significance can best be addressed locally, all according to the impact of the subject upon the proportionate interests of those subjected to it.

This raises the question of the propriety of democracy, a process that by definition opts not for the best and certainly not for the worst by intention, but for the mean between. Under circumstances of dire distress, democracy cannot be relied upon either to keep us out of trouble or to deliver us from extinction. Only intelligence, i.e. thinking the right thoughts and taking the right actions, can enhance our prospects of survival.

Therefore, by what means can the populace benefit from exceptional minds and exceptional skills where democracy dictates government by the average minds of average citizens?

Question Authority

Every time that we make a decision with respect to the exercise of authority we need first to ask ourselves the most fundamental questions: who is affected by the outcome, what are the relative merits of their special interests, whose interests are greater, and why? The means by which decisions are made need to be consistent with the impact of such decisions. This is the problem central to democracy that must be constantly detected, directed and corrected. We need at each turn of events to reiterate the question that arose earlier: towards what ends are we intending?

The challenge is to appropriately match forms of organization with forms of enterprise. Thus plutocracy (government by the rich) might be the organization appropriate to those endeavours concerned with balancing budgets, financial planning, investment, etc.; timocracy (many with value, loosely translated to mean virtue) would be the preferred mode for the administration of (say) education or healthcare; aristocracy (rule of the best) to govern technological enterprise; democracy (many without virtue) to over-view for potential misfortune; and meritocracy (the philosopher-king) for the guiding of spiritual matters and overall direction.

To the extent that democracy enables people to gravitate towards the kind of work for which they are best suited, this fitting process occurs informally. It is only in emergencies therefore that responding with the most fitting form of authority may be critical to averting a greater emergency. It thereby behoves a community to position its contingency resources in a state of readiness where they may be activated proportionate to the demands of the occasion.

Change for the Good

We return to Plato's understanding that virtue is the correct standard by which to classify governments, which idea raises the question as to how societies can redesign themselves based upon ever-improving models of excellence.

How can man maintain the practice of virtue as a system of behaviour whereby vice is not sanctioned? This may sound like the system we have in place and indeed is the basis of our system, but what has happened to our value system is a quantum warp in values in favour of money and power at the expense of all other values. How, we may ask, can we get onto a more honourable path?

The outstanding need with respect to authorities at all scales is for the placement of critical limitations upon their capacities to corrupt virtuous interests, and the continuity of society's abilities to enforce compliance with such limitations under the law. The question of appropriate scale is germane because sovereignty or domain necessarily shifts in the winds of time, as from local to national to international, in order to offset corresponding shifts in the scale of corruption of virtuous public interests.

Logically, if the US was unequivocally convinced of the efficacy of democracy, it would channel its energies into worldwide democratic organizations like the United Nations, sign international treaties aimed at safeguarding natural resources, take a neutral position with respect to mitigating disputes between other nations, and take its place in the family of nations proportionate to its population and the will of the majority. Hopefully in time it will, of its own free will or because there will be no feasible or thinkable alternative.

Democracy does not necessarily make good decisions, but it tends to avert the worst. This is what it is good at doing, and this is what is should be directed to do.

Historically politics has not taken its cues from ethics but rather from expediency. The main thrust of political activity has been the establishment of authority for the purpose of implementing coveted agendas. This focus upon securing and maintaining the power of decision-making by an elite few that are bolstered by military and economic power runs counter to the more general interest of serving the best interests of society as the highest purpose of political activity, giving rise to the question as to when an elite becomes effete, one answer to which is when it becomes corrupted.

The maintenance of political parties that polarize distinctions between competitive ideas rather than attempting to reconcile them is counter-productive. When a man-made organization becomes an end in itself rather than a means to the betterment of others, then it is time to change its constitution and behaviour.

As to the question whether democracy and excellence are mutually exclusive, the answer hangs on our definition of excellence and the extent that we use, or rather abuse democracy by insisting upon its governance in all matters of consequence – a totalitarian disposition.

If there is a single quality that a leader of men should have, it should be that he be humane.

146 Leadership: Advancing Change into the Unknown

'Some are born great, some achieve greatness, and some have greatness thrust upon 'em.' Shakespeare seems to have articulated a truism that has survived long past his *Twelfth Night!*

Today the term 'great' is banded about as an involuntary reflex to signify acknowledgement of something, anything, without further elaboration; an inarticulate exclamation or 'joker' that invites the listener to fill in an absent meaning with significance of his own choosing! So what is so great about greatness; indeed in the context used above, what does great mean? Einstein's likely conclusion (abstracted from Section 1, page 1) would suggest 'it all depends'!

The Concept of Leadership

Some thing or event may be said to lead another if it precedes other things or events in time. Such precedence may be construed to have potential for influencing later events. Thus, in the realm of human affairs, what it and all depend upon, and leadership opportunities offer, are degrees of relative influence upon lives and the course of events through the exercise of authority. As we have already established, authority may take many forms.

Insofar as such authority may be exercised through exemplary conduct or brute force, leadership may be positive or negative. Over an extended time, like the pendulum, on average the effects of leadership would appear to swing towards a position of neutrality.

The Power to Cause Detrimental Effects

Whereas gains through good example are generally registered incrementally, the effects of single decisions authorizing the use of excessive force are potentially catastrophic. Therein lies the fallibility of leadership. Therefore, on face value, it is better to waver in the direction of neutrality than to opt for excessive force. History is replete with examples of excessive authoritarianism which, when it extends beyond acceptable limits, requires overreaching counter authority to neutralize.

Power to Advance Beneficial Progress

When we invest leaders with authority, we do so in the clear and certain knowledge that we are extending them a large measure of trust in their capacity to guide us to better, more secure futures. At the outset we often entertain great expectations. Invariably such expectations are tempered in the passage of time until such time as we are able to look back impassionately and, upon occasion, be entertained by the naivety of our earlier thoughts.

The question of trust is pertinent because, as we have established, except in cases of emergency, we prefer to extend trust incrementally, depending on how well prior trust has been justified by events.

The Need for Reason

The extension of trust is a function of judgement. Such judgement requires justification on the merits of the situation if the consequences of choice could be significantly different in the event that a different decision were to be made. Judgement is coloured by whether one is guided more by one's emotions or by reason, whether one is interested in short- or longer-term benefits as returns on an investment in a decision and whether one is directly or indirectly rewarded.

Qualifications for Leadership

The question naturally arises what qualifications befit a person for leadership. We have already answered this question with respect to the added security afforded by prior justified trust. In areas of specialization and professionalism we do not hesitate to seek the best-qualified help.

Ironically in politics, and most particularly in democratic political environments, there are no prerequisites for the assumption of leadership roles save the necessity to be elected by a majority! Therefore, all that an aspiring politician needs to do in order to assume responsibilities disproportionate to his single vote is to persuade others that he is the best candidate for a particular elective office. And all that is required to achieve this is a preponderance of influence. And what better leverage tool for the promotion of influence is there than money? And how much greater advantage may be obtained if the ways and means used to gain influence are corrupt and unacceptable to opponents as their means to their ends? And how can such conduct be construed to be anything less than contemptible? And how, or rather why should we be so surprised or upset when individuals with few or no qualifications to lead men are elevated to such positions of authority on the basis of meagre qualifications?

Thus the innocent and confused masses upon occasion extend their trust to an unknown individual to lead them like lemmings into an unknown future.

Clues for the Clueless

On the side of caution and common sense, are there any litmus tests for responsible leadership qualities? A few character traits may be read by an electorate that can reinforce confidence in their decisions to elect certain leaders rather than others. Preparedness, whether by training or experience, is a signal of leadership potential. An ability to assume responsibility, even where it has been demonstrated in a less influential arena, is good cause for a modicum of confidence. A demonstrated ability to relate to people well bodes well. Self-discipline is a hallmark of prospective leaders. An ability to listen and learn are worthy attributes. Positive motivation and tenacity, diligence in caring for the concerns of others, respect for human dignity and a sense of justice all portend for more of the same. An understanding of human qualities, moral courage, endurance in the face of adversity, initiative, creativity, flexibility, receptivity to the good ideas of others, and inspirational qualities to lead men in common cause are all admirable and desirable traits. All these attributes are as patches in a quilt intended to secure people against the harsh realities of their environment. To the extent that such qualities suggest better prospects for such coverage than alternatives offered, they afford grounds for optimism.

After-Effects

The effect of good leadership is to deliver rewards more promising than were otherwise probable. Leadership is neutral. By definition it neither promises benefits or costs, but in general it delivers both. The question after the fact always seems to be an economic one: were the benefits worth the costs?

147 Freedom: Relative Absence of Restrictive Relations

The term freedom is one of the most freely bandied ideas in the English language. It is so free that one often has difficulty understanding what is meant when the term is used. This is because the idea of freedom hinges upon contingencies that are less freely expressed. The 'absentee' is independent or free from expression and therefore from comprehension.

Freedom from What?

The idea freedom suggests to us that we are uninhibited and unrestrained from doing anything that we wish. It is only when we try to apply that idea in practice that we discover the idea to be conditional, constrained, limited within bounds. So the use of the term denotes a euphoric dream-ideal of implicit independence of action. We are free to jump up, but not free from falling down. So the first lesson learned is to ask: of what do we presume or aspire to be free of?

Of the many freedoms cited by King John in the Magna Carta, the first, that 'the English church shall be free' falls short of stating what it shall be free of; the second granted and confirmed 'the freedom of elections' without stipulating who the electors were to be; and the third, 'We have granted to all freemen of our kingdom, for us and our heirs forever, all the underwritten liberties, to be had and held by them and their heirs, of us and our heirs forever', grants certain rights to certain persons without defining who freemen are, but rather infers through the recitation of entitlements that freedom is to be allocated according to the established hierarchy of privilege, to the privileged; everyone intended to gain some advantages as rights, as in the case 'To no one will we sell, to no one will we refuse or delay, right or justice.' We are led by this and other statements to infer that those rights that pertain were intended to apply to all free men or freemen equally, distinguishing them from slaves, vassals and convicted felons.

All this was more than 300 years before King Henry VIII freed the English church from papal jurisdiction.

Freedom from Want

Henry VIII serves well to illustrate how missing qualifications come to be interpreted when they are not predetermined. A man of many talents, all of which gave him great ambitions for personal fulfilment, Henry was the quintessential egotist. His unstated freedom from wants dictum was directed in the interests of the first person singular: I want.

All his actions can be seen in retrospect to have been taken in support of his presumptions of best personal interest. Thus, in 1521, when he lent support to the Pope's opposition to the breakaway protestant Martin Luther, he did so because it was politically expedient (for which he was bestowed the title 'Defender of the Faith'). Just twelve years later, in support of his interests in expanding his political power base at home through appeasing a growing popular anti-ecclesiastic sentiment in England (through which he gained parliamentary support); in seeing a clear path to legitimizing his interests in a line of succession with a survivable male heir (he had lost two sons); and in recognizing that, through the dissolution of the monasteries, all the wealth of the church would pass to the crown, he had no difficulty in effecting a separation from Rome and adopting the Protestant faith for the expedient price of a few Catholic martyrs.

No longer constrained by Rome he secured a divorce from Catherine of Aragon and married Anne Boleyn. A pragmatic blueprint for successful opportunism had been forged that enabled him to advance his interests through the devices of marriage, divorce and execution for the balance of his reign. His stature as a monarch rests upon the coincidence of his personal interests with those of the state. The interests of the populace were on the ascendancy. While he is renowned for his break from Catholicism that enabled the Reformation, his efforts to unify the British Isles, his parliamentary enlightenment and his development of the navy were essential to subsequent colonization and other exploits of Elizabethan England. Under Henry, freedom and liberalism were synonymous. Freedom from want became everybody's want.

Freedom of Relations

We have come a long way in securing freedoms for men but clearly the ghost of Henry's egocentric mentality lingers and often dominates our minds. The positive side of the balance sheet is our interest in goods: good things and good relations, however they may be defined on occasions. No one would deny the potential for beneficial effects that flows from such opportunities. What is less clear is the divide between what constitutes the good of one person and a detriment to another, and under what conditions; but in general we are most interested in securing freedom for ourselves, less so for those with whom we identify closely and still less for all mankind.

What we would hope for would be many instances of correspondence between the definitions of good of many individuals, upon which we may build consensus. On subjects like the necessity for clean air and fresh water one would think that there would be little opposition, but if the by-products of a person's livelihood have deleterious effects, and if he has no access to alternative opportunities, then he is going to defend his actions militantly. It is therefore the freedom to form beneficial relations between people that is precious, but always subject to social restraints and sanctions if they should be found not to be so. Insofar as relations are reciprocal, there is an obligation on the part of collective man to offer relief in some form to those that are indisposed. Viability of life is conditional. If such conditions are undermined by dark portents, the causes require to be disclosed and remedied.

So we advance freedom of positive or innocuous relations as a general rule while curtailing freedom to cause detriment to others. How we do this, at macro or micro scales, is subject to political will, law or military action, always understanding that all problems are incremental and that some problems are more insidious than others since they take effect stealthily. Today we hold ourselves free to play out our evil fantasies in games and 'entertainment', unaware of the stealthy influence that these may have upon our minds.

Freedom from Relations

The negative side of the balance sheet is the perception of negative impacts due to overbearing burdens of political will, law or military action. Each of us views these from a different perspective, understanding that the inertia caused by such a large body of constraining force undermines our own abilities to exercise our personal powers for the good. Political doctrines close minds to possibilities; they corral people into conformity. Conformity denies freedom from conformity. Freedom is anathema for those who don't have the courage to embrace it, much as totalitarianism revolts the liberal (free) mind.

Democracy itself at times appears to be, if not a pollutant then a dilutant that detracts in equal measure from the prospects and opportunities to perform exemplary good work as it imposes constraints upon negative conduct. We are inundated by the imposition of obligations, demands, intimidations, threats of retribution, unsolicited communications; indeed, all the freedoms of others tend to interfere with our own interests in quietly minding our own business.

We are in an age of specialization in which we build territorial boundaries around what we do. In doing so we lose connections with others and in large measure the stimulation that enables new points of view, new relationships and new Leonardos to emerge. In order to empower creative men and women to flourish they must be free to explore and discover.

What each of us needs is an override switch to stop the earth when we want to get off; a respite from interference beyond the purview of the public domain; a sanctuary within a sanctuary. Such prospects do not simply reside in our minds, they 'are' our minds.

Men cannot be truly free until they are secure, and cannot be truly secure until they are free.

148 Dependence: Reliance upon Relations

As demonstrated by the 'Peking Butterfly' effect, all relations are mutual and nothing is spared connections. In this sense there is no true independence, merely mutuality of existence.

If, as we have earlier asserted, all there is is all there is, in the macrocosm of 'all-there-is', all there are are relations.

So it would seem to be elementary that we should concentrate our minds, and those of our children, upon understanding the nature of relations, and on nurturing those that are beneficial.

Symbols of Symbiosis

In real life, dependencies are formed between partners for their mutual benefit. As to which partner benefits the greater and which may be wholly dependent upon the other, whether two can reciprocate on an equal basis or whether one simply accommodates and tolerates another, is a function of circumstance and perception. But it is true to say that all living forms form dependencies directly or indirectly upon other living or formerly living organisms, each of which forms other dependencies on other living or formerly living organisms, etc. The loose definition of symbiosis applies to coexistence or cohabitation, the 'sym' referring to a combination of forms, the 'biosis' to a mode of living.

While all this goes on without so much as an invitation, and certainly without consciousness as we have defined it here, it nevertheless can be shown to be beneficial to one or both parties and generally at no great cost to either. Collaboration between dissimilar organisms offers each access to that which it would otherwise not have.

Reliance upon the Reliable

More detachedly, we seek sources of experience that are time-tried and proven to be reliable. Gravity is one such experience. Imagine a state whereby gravity acted as an alternating electrical current; we would never know with any degree of confidence which way was up! Well it does, kind of! All the bodies in the universe pull each of us towards their centres. It is only the close proximity and relative masses of each of us to the earth that adheres us to 'here'. Upon this we can rely. Our lives and existences are bound by such boundaries. We thrive because we are so thriven or grasped.

Why then is it that upstart man deigns to deny the realities of interdependency? Why should he so pretentiously assert his independence? How is that he assails the unassailable truth that we are dependent upon relations with all else? He is testy. He is testing the system. He has not yet risen to a state of trust in the state of 'all-there-is', there is something missing: trust in his own dependability upon himself, or more specifically upon his own state of mind.

Building Dependence upon Reliable Social Relations

In the interests of stabilizing his erratic behaviour, man's obvious option lies in committing to a system of mutually reliable relations with those who share his dilemma. This can only be achieved on a basis of equanimity.

There are only three generic types of relationship. We have seen how forces sail through life with no obvious relationship, as the moon does to, say, Halley's Comet; we have seen how they converge upon each other deflecting the course of each and dissipating their energies; or they can act in-line, either on a collision course towards mutual destruction, or in the same direction whereby they combine their energies and masses to gain momentum and potential. Well, that is the end of the physics lesson. Let us entreat unity of purpose in all matters of mutual dependence. In so doing, in Milton's words, 'They also serve who only stand and wait'.

149 Discipline: Internal and External Control

Discipline is a state of order by design. Some are born disciplined, some acquire discipline and some have discipline thrust upon them. And then there is objective discipline.

Born Discipline

At birth we have certain proclivities and propensities built-in as predispositions towards environmental forces. As to whether there is a 'discipline gene' or not we shall leave for others to determine; suffice to say we can and cannot do certain things naturally.

To the extent that we are all born into a world inhabited by discipline, how does this influence our lives? Whether discipline is an advantage or a liability depends largely upon what one is doing or trying to do. Generally speaking, discipline is better applied to the service of repetition than to creativity, an endeavour that makes claims on the exceptional rather than upon precedence.

If one needs more discipline than one has, one has but one option, to acquire it!

Acquired Discipline

If we desire or are required to do something that we have never done before, we need to try. The practice of trying cultivates abilities. That is one of the principal lessons learned when growing up, an article of faith in formal education. If one concentrates upon a goal to the exclusion of all alternatives, the likelihood of achieving that goal is immeasurably enhanced. To that end, even the goal of acquiring discipline requires discipline.

Imposed Discipline

Discipline is the means by which society exercises control over individuals for purposes that range the whole spectrum of possibilities from the well-intentioned to the malevolent, with corresponding beneficial to disastrous results. Beware of imposing imposters and disciples of discipline!

With regard to discipline, the problems of society are identical to those of the individual in this regard: how to bring order to bear in a realm where entropy always enjoys the advantage.

Just Discipline

A state of order, as we have discussed, exists in any arrangement due to relations. As to whether such order is static or in movement does not detract from the fact of its existence. We use the term discipline to convey a state of being or beings that exhibit steadfast characteristics.

In the final analysis mankind's struggle is not simply between reason and the emotions, or between right or wrong, but in equal measure between discipline and the absence of discipline, as mind bears potential for what matters.

150 Consent: Assent to Advice

Underlying the necessity to make judgements and take decisions that lead to action is the need to determine the appropriate weight that should be accorded to principle vis-à-vis expedience. Nowhere is this more evident than in the field of politics.

Insofar as political process compromises principle by its nature, accommodating fragments of opposing positions in the interest of forging a majority consensus, the individual who stands on principle on an issue, particularly when to do so contradicts the positions of those that most closely correspond to his general political persuasions, is placing himself at risk of losing their loyalties.

Compound Judgements

Political judgements are made in the context of an understanding that inherent in the nature of any decision is the necessity to take into account the effect of a particular judgement upon the relations between politicians, fully understanding that they are mutually dependent upon each other for support of their most favoured topics in successive debates. That is political reality!

From the above we can now clearly understand that inherent in all such decision-making lies a spring-loaded potential conflict of interest!

Factional Fraternity

To the extent that one's general views tend to be inclined to favour one pole of the political spectrum, one's tendency is to make judgements and cast votes automatically by default. Principle, insofar as it applies to a particular decision, derives more from such a preposition than it does from the merits of the issue under consideration. Those so afflicted are in effect blinded by their prejudice and, to a large degree, predictable as to how they will cast their votes. This is the obvious downside of the political party system that exerts undue pressure upon party members to vote in a block in order that they may prevail, even in the face of clear evidence that a particular decision will not lead to such beneficial action as it may otherwise have.

Independence

Against this background one may then entertain the prospect and consequences of making judgements in clear conscience solely upon the merits of each case. Here the exercise of scrupulous honesty pays dividends, not only because one can address each subject strictly upon the evidence of causes and probable effects, but because in doing so one gains respect from those that choose to divide their commitments between the merits of the issues and party allegiance.

Compulsion

Whether one feels compelled to make a judgement in clear conscience or under duress, it is in this context that consent to advice is weighed and displayed.

151 Promise: Commitment to Intent

The purpose of raising the subject of promise here is to examine what a promise really is, what its parameters are and under what circumstances it loses its utility as intended and becomes an instrument of persecution.

Like so many ideas that we have addressed thus far, the word ‘promise’ is fraught with nuances that tend to lead one away from the obvious and into the realm of uncertainty. The first premise in considering the notion of a promise must be that a promise be given freely rather than under duress, pain of suffering or with ulterior motivation. The circumstances surrounding the giving of a promise are presumed to be known to the promissor.

A Promise is a Promise

A promise is a declaration to perform or not to perform a particular act. It is a pledge, as of loyalty, to honour one’s expressed undertaking, which pledge gives rise to expectations in others. It is assurance that, to the extent that one is able, such a pledge will be honoured.

But an affirmation is not a delivery, still less a guaranty; it is rather a declaration of intent to act in a certain way to the best of one’s ability. The parameters that reinforce the promise are those that are within the control of the promissor, those that are not may not.

Pro-Misery

When, one might ask, is a promise not a promise? The obvious answer is when it is false, that is, when a promissory word is not in consonanance with a concealed intent; when it is fraudulently given for the purpose of stealing advantage, i.e. when it is deceit.

The Promissory Note

All written contracts for the exchange of goods and services or for the repayment of debts are promissory notes. By virtue of their being written they become hard evidence in support of claims of breaches or positions with respect to disagreements over the terms of promises made.

The Promissory Oath

But there is another set of circumstances where the ideas of fraud and deceit do not come readily to mind, even when the promise is used against the promissor, and this is when the promise is institutionalized in the form of an oath.

Let us look for a moment at the oath. One’s resolve is a private matter, one can choose or not to share one’s resolves and when one does, that might be considered to be a declaration of intent without prejudice. But there is a more formal declaration, generally made before witnesses, that moves a commitment from the private to the public sector, for example a promissory pledge to speak and act in accordance with one’s understanding of the truth. We have seen how tenuous truth is, and how ephemeral. How can it be argued and expected that, by pledging an oath, the truth can be any more clear or certain? This is the so-called oath, not to be confused with an irreverent or profane declaration of the same name.

But can we so clearly separate reverence from irreverence, particularly because we use the same word to describe both? We rely upon the context in which the word is used to signal the intended meaning of the word. Not only is the context subject to differing interpretations, but so is the word itself. Does the wilful making of this disparity in meaning not create of itself deceit with regard to the validity, indeed the virtue of any such oath? And how is the oath used in practice? It is used ostensibly to cast a more credible impression of honesty upon an otherwise uncertain, perhaps gullible audience on the one hand, and as an umbrella to hide under when one elects not to tell the truth as one understands it. In a nutshell, it casts an unwarranted cloud over the notion of honesty for, by implication, one who does not so swear by oath is deemed less trustworthy than one who does, and is thereby less likely to tell the truth and, by inference, his word is less likely to be true.

So why would anyone elect not to pledge an oath when requested? People seldom do, and that seldomness is the proof of the pudding, that the person of honourable intent and the chameleonic pragmatist or dubious character each pledge their formal oaths equally willingly because such pledges carry with them the weight of conventional prejudice. And how does the law distinguish between the two? By placing each under penalty of perjury if they should knowingly deceive, mislead or bear false witness to events and knowledge, as they know it. The state sets the individuals up in a state of fear of social consequences if they do not hold to their oaths. And what happens to those who do not so pledge their oaths? They are likely to be held in contempt for declining to honour the time-honoured, if questionable, practice of pledging.

So the cowardly cower and pledge the oath, and the brave are vanquished by the cowardly but powerful. Isn't that just how it works in real life? And where does the conscience fit into this puzzling situation? It is either set aside for convenience, perjured by the act of pledging the oath, or becomes the basis on which society segregates and penalizes those whose consciences dictate that their native honesty overrides their socially conventional dishonesty where they are in conflict.

The oath is a fundamental flaw in human relations for it attempts to guarantee what by its very nature is temporal, uncertain, changeable and subjective: truth, the very sanctity of which is held to be the reason for requiring the pledging of an oath in the first case.

The Secret Side of the Oath

What are the courts seeking when they invoke the oath 'Do you swear to tell the truth, the whole truth and nothing but the truth, so help me God?' Such a question is unintelligible. Neither absolute truth nor God are accessible to verification and therefore cannot rationally be admitted as evidence. As long as courts predicate their judgements on incantation (the use of words to obscure rather than illuminate) and on absolutism, justice is thereby categorically denied. The concepts absolute truth and God have no place in the oath, so they must be there by default or worse, to intimidate the individual. Indeed the sole purpose of the oath appears to be to create a pretext for punitive action under the authority of the court if the court does not like what it hears. The very requirement of oath-taking admits to a weakness in the court's position vis-à-vis individuals whereby it is seeking the authority or at least the concurrence of the individual to rest its own authority on. Without that grant of authority the court becomes authoritarian and penalizes the individual for not taking the oath, perhaps as a matter of conscience, where the court does not have conscience and can therefore act unconscionably. The court may now, in the absence of a conscience of its own, hold the individual in contempt of court, and act in a contemptuous manner to restrict the liberty of the individual to exercise his constitutional right to free speech, criminalizing the individual who may have done nothing and said nothing. Is it an acceptable public policy to put citizens in jail when they do nothing and say nothing? What kind of institutional inadequacies are being covered up by such official behaviour? Where is the boundary of totalitarianism?

What we have been arguing is the logical absurdity of the oath, and the necessity to acquiesce to its social purposes.

Courting Disaster

We reflect again upon the experience of Alice in Wonderland wherein the court declares its judgement and sentence and then proceeds to hear the case. This we might think ludicrous on face value, but this is exactly what the courts are asking litigants and witnesses to do when they insist upon the swearing of oaths prior to participation in a case. The reason that a court suspends sentencing until the end of the proceeding is to enable it to take into account the evidence and circumstances of the case. This is exactly what the individual is being asked 'not' to do by the requirement to swear an oath. He his being asked to foreclose his options with respect to what he believed to be the truth prior to hearing the evidence and taking the circumstances into account. What the courts are doing in requiring the swearing of oaths is intimidating and corrupting the plaintiff, defendant and witnesses, and wilfully obfuscating the truth by overburdening jurists with unnecessary bombast and inadmissible 'com-promise'.

The Alliance

On a more promising note, the alliance between Great Britain, France and Poland to join in common cause in the event of an attack upon any one of them by another nation was arguably the most significant and far-reaching promise ever honoured in history. On September 1, 1939 Britain upheld its commitment to Poland by issuing an ultimatum to Germany just hours after Germany attacked Poland. Two days later Britain and France declared war on Germany for the second time in one quarter of a century. The alternative to this courageous action does not bear contemplation.

152 Agreement: Unity of Persuasion

Agreements occur when people's 'thises', 'these' or 'theses' correspond. If they cannot reconcile their notions in the present tense, then they are greatly disadvantaged in their abilities to reconcile their futures. Therefore, in order to proceed with a mutually beneficial relationship, the need exists for a common understanding of what each 'this' refers to. We can quickly appreciate that in the absence of a common language, while 'this' may be 'this' conceptually for two people, the sounds and the literations of their descriptions of it are different, thereby creating a basis for misunderstanding that 'this' may not be 'this'! This dilemma leads us back to our problematic 'maybe', and uncertainty.

Sign Here

So what does it take to make the simplest form of agreement? A sign. Any sign that 'sign-ifies' that two people are thinking, believing, acting or planning to act in accord. Even an action by one in the interests of another is enough to tell the second that there is a force for unity of persuasion at work. Two people working towards the same end form an agreement in principle. From this simple form of common understanding there arise agreements as complicated as man can conceive. The trouble with complications is that they introduce too many contingent 'maybes', any one of which, if it should arise, could unhitch the parties from their agreement, or worse, lead them into odious disputes.

Beware of the Sign

A signature upon a written agreement transforms it into a special kind of promise: not just a promise between two well-intentioned people, but a contract that can be interpreted or misinterpreted, with good or ill intentions, for better or worse by others!

The best agreement in principle is a nod, a handshake or similar confirmation or action that overtly confirms for both parties that there is a shared understanding. It is the maintenance of principles that hold such agreements in good standing.

It is only when the stakes are high that the need exists for a written contractual agreement, so that, in the event of misunderstandings or the failure of one party to perform, the disadvantaged party may have a legitimate basis for attempting to recover from his losses.

Yes!

The premise behind an agreement is that two or more parties subscribe to an idea by signifying 'yes'. 'Yes' means 'this' is 'this' in their hearts and minds. There is an exchange of common understandings.

'Yes' has a special relation to 'truth' and we know how problematic 'truth' is! There are almost as many linguistic paths back to the roots of the word 'yes' as there are to sources of truth. At their confluence we find the common idea 'this is so'. Let us take a quick look at one such path:

Ita (the Latin adverb for the affirmative) = (is) 'so', 'thus'.

Est (the third person singular present indicative of the Latin verb *esse*, to be) = 'is'. (Used as a prefix in Modern English, as in 'est'-imate to 'est'-ablish an 'est'-ate of truth, etc.; and as a suffix to designate the high-'est' and b-'est' standing in fact or belief that something 'is' so.)

Sic (the Latin adverb) = 'thus' or 'intentionally so'.

In their combined senses they lead us to concurrence, 'so thus-is-thus, intentionally so': **IES**!

153 Fulfilment: Satisfaction of Needs

Each of us according to appetite, enterprise and commitment harbours and sustains needs to satisfy deep-rooted urges that run silently within the egocentric self. This is no reference to over-reaching greed, the grossing out on netting in more than necessity prescribes. We refer to the call to unspelled duty to actuate an appropriate response to needs, and to the measure of pleasure that returns. This is the spirit of the ego that, 'like any other sovereign, god, king or thing, is going to radiate its authority and through that radiation distribute its influential parts'; for which the ego is justly rewarded.

The Outgoing

Upon occasion, as a tide of energy with good intent, man commits himself to service, to give and not to count the cost, to fight and not to heed the wounds: the urge to merge into union and to emerge at an end that extends beyond mere compensation.

Tranquility born of spent energy stays the high tide in ambient stillness, resting before receding. Having given its fill in full it is fulfilled.

The ebbing tide circulates much among the many, mixing all with all else. The mid-tide swirl drags away all signs of lassitude as all is excited into motion. This is the sweeping moment of action when waste is abated following the feeding of needing.

The low-tide dillies, as if in a quandary, equivocating between the pull and the push, until nature's secret force equilibriates to inevit a turn and return that is its destiny. We are all in service to all.

The Incoming

As refreshment of nutrient feelings that replenish that which is spent, tides shallow-fill the lower basins of the inner sanctum with satisfaction. It is the satisfaction that is born of having satisfied. The sense of having done something that was worth the cost in full: a good day's work is repaid in full 'fill-fuelment'. There cannot be an incoming without an outgoing.

Fulfilment is enriched by degree according to one's service to the one.

Man compensates man with goods, services and token values. Nature rewards him with knowledge for acting in consonance with her purposes; the knowledge of having acted correctly, honourably in relation to nature's needs to continue doing what she does so well.

Capacity

One cannot consider fulfilment independent of capacity. As the word suggests, to full-fill and to fill full speak of two quantities, that which is filled and that which fills. Without acquaintance with the nature of each we are at a loss to understand the meaning of fulfilment. The capacity of a human being to realize its goals and, in the context under discussion, to feel private satisfaction, varies greatly. We cannot therefore assume that all are uplifted by their behaviour or feel rewarded to the extent of their deserts or beyond their expectations, nor that they so aspire.

Conversely, we cannot deny the capacity of anyone to so feel under favourable circumstances or motivation. Anyone is capable of saving anyone else at any uncertain time in the future.

Content-ment

There is also the intangible aspect of the quality of that satisfaction, i.e. whether it is self-serving indulgence, distinguished from nourishment, or a bath in glorious reflection for having served others well. The question lingers, with what do we each need to be filled to feel fulfilled? Perhaps Karl Marx offered as good an answer to this question as any with respect to our individual capacities for giving and receiving: 'From each according to his abilities, to each according to his needs.'

154 Compromise: The Act of Reconciliation

Compromise should not be viewed as the domination of one idea over another but rather as the reconciliation of forces proportionate to their contributions. When we experience the confluence of forces it may be helpful to formally represent them as geometric expressions in order that we may simplify and thereby elucidate appearances of complexity.

Life is not composed of relations that are obviously simple to represent. Complexity abounds. What is needed is a mechanism of reduction. Such a means is analysis. We take apart to comprehend each part. We examine the parts for what they can and cannot do, and we give them separate identities. From such understandings we are able to reconstruct or create new combinations of relatedness, to synthesize new ways of doing. We combine relations with relations and in so doing build new orders of complexity. With each compound addition we tend to reduce the performance of the parts in order to give new attributes to the emerging whole. It is therefore incumbent upon us to refer to our intentions for the performance of the whole before detracting from the performance of its parts.

The Archetype Compromise

The most cogent representation of what does or could embody the 'best' performance of parts is confronted by the necessity to yield to time and other competing interests in support of reconciliation. Two 'bests' may not be mutually compatible. In such cases establishing the resolution of their conflict becomes of higher priority than the optimal performance of one or the other part of the system. Appearances suggest that the principle of giving, as of integrity from the part to the whole, is a necessary precondition to the development of more complex organizations. This is so in aggregation, whether human or otherwise. The issue of principle vs expedience reduces to scale of operation with regard to the parts vis-à-vis the whole: the smaller vs the greater interests. Politics is the profession and practice of hierarchical social compromise. It is not that compromise is good or bad, but rather that it is inevitable as two or more forces resolve to become one. Goodness or badness arises only as a function of individual perception.

The Compound Compromise

To recognize the cumulative effect of compromise, let us examine the respective contributions of independent systems to the integrity of the whole, and enquire into the nature of the compromises involved. To identify the essential nature of each system one must ask the questions what is it doing, what is its potential; or, in the case where it is in development, what is probable, what is possible and what is intended? Consider the exterior envelope of a building. Here the claims of the individual subsystems compete to occupy the same space or plane at the same time to satisfy discrete criteria for their own functional integrity. To establish the degree to which each system may be compromised we ask questions concerning its criticality.

Uncompromising Systems Integrity:

The physical superstructure of a building is fundamental and has precedence in design priority over everything else. Loss of structural integrity translates to loss of building. While the structure may be offset inside or outside the plane of the enclosing envelope, all systems depend upon it for their own integrity. Almost as critical to the utilization of any building are its moisture-barrier and wind-shield systems. Water damage represents the single most costly type of building failure after structural collapse.

Relative Systems Integrity:

Heating, ventilating and air conditioning, thermal insulation, lighting, acoustical considerations and aesthetics all have a strong bearing upon levels of physical and mental comfort; the better they are, the better we feel! But man is adaptable within broad ranges of tolerance.

Intermittent Systems Integrity:

Access, egress and building security are discretionary systems under control of management and subject to the criticality of the building use.

Elastic Systems Integrity:

Expenditures of time, money and other resources are variable subject to availability and the forces of the greater environment of which the building whole is only a part.

An appearance of a unified system of systems emerges as we recognize that the wrap of a building is a complex security blanket protecting life and property within. Affording this protection is the bottom line in all matters of compromise.

Many man-made artifices are compound mixes that compromise individual disciplines to a greater or lesser extent. We return to the subject of priority in order, and thence to priority in time to determine what governs what in the order of orders.

Compromise as Process

If compromise is intended, it is achieved through the operation of planning, judgement, decision and action, all of which we have earlier discussed. Whether intended or not, compromise naturally occurs proportionate to the strengths and directions of contributory forces. This principle we have discussed under the general topic of 'Force: Energy Acting in a Cause'. The incidence of intention places the burden of initiative upon the intender as the primary cause of the application of force of a predetermined value in a predetermined direction in order to assure that the intention transforms a possibility into a certainty.

The Reality Check

There remains the need to test and verify that the performance of the new whole resulting from compromise is in accordance with expectations. If not, further compromise may be induced in order to establish improvements in performance. In this way compromise may be viewed as a process of gaining benefits in increments rather than as concession, always understanding that where many interests are at stake, the greater the extent that benefits are distributed, the greater the integrity of the whole. Our very acceptance in society is the product of such compromise. Subscription to rules of morality (for example) is a necessary test to be passed in order to gain social acceptance and access to the 'good' life.

The conclusion that we are inevitably drawn towards in all considerations of compromise is the necessity for getting one's sequence of integrities in order, of which the most fundamental and profound is personal integrity.

155 Gain: The Accrual of Quantity

Gain is accumulation in quantity. Even gain in quality is an accumulation in quantity. When we examine what 'gain' means, we find ourselves confronted by a corresponding idea: 'loss'. We must ask fundamental questions like: where do gains come from? Is it absolutely necessary to take in order to receive? Is the idea 'transfer' an essential aspect of gain? Must we give in order to receive? Is such giving involuntary when it is not voluntary? Is this Christian ethic simply a physical principle of nature dressed in 'Habit: The Mantle of Disciples and Sloths'?

In identifying with gain we acknowledge positive interest. The converse is, however, also true; gain in pain meets with our rapt attention. So some form of discrimination suggests itself whereby we need to segregate beneficial from inauspicious gains, advancing the former and remitting the latter.

Means

How we plan for the accumulation of positive gains rests in the first instance upon defining who 'we' are. If we construe 'we' to be each one of us separately, then we are on our own. If we interpret 'we' to include others than ourselves, then we necessarily qualify and modify our planning to that end. Most of us find our interests to be vested in the 'we' of ourselves and the 'we' of a group in some mix that is of itself in flux over time.

Ends

On first blush more of anything that is not negative is positive, neutrality harbouring potential for good in the absence of being overtly negative. Experience has taught us well that a little gain-planning holds us in good stead.

We have developed sophisticated ways to accrue neutral resources, of which money is the most obvious. Money can be applied to good purposes or otherwise. We bank it to provide for contingencies of need, as security a-gain-st uncertainty and in the cause of wellbeing. In all such cases, neutral resources are an asset. Superficially gain in pain equates with loss in security. But when we take into account the effort required to gain benefits, the cost is often not without pain.

Kinds

The forms that positive gain take are of great variety. In general gains may be distinguishable by class as benefit, security, improvement, welfare, fulfilment, education, commerce, value, mass (quantity), happiness, etc., each of which has, by corollary, its antithesis.

Unkinds

Costs must be taken into account in securing gains. If such costs are neutral, then there is no detrimental effect to securing such gains, but if the cost is great, particularly to others, then consideration of their interests is due, as if they were one's own, by which we are reminded that we are somehow bound together.

What we observe is a kind of unkindness that often attends the procurement of positive gains, as of taking from others at their expense. This is the distinction that we acknowledged with respect to entertaining positive vs the negative experience, and 'we' as individuals vs 'we' in the plural.

Time Out

What we discover is that volition cannot be divorced from the affections or disaffections, and that positive gain is only one side of an equation that in principle tends to return to a state of equilibrium. Plus tends to weigh-in equally against minus, zero-gain translating to zero-loss. In terms of negation, less is more; a more positive position by virtue of being closer to equilibrium.

The Concept of Profit

In order for motion to be maintained according to the principle of inertia, force and direction require to be held constant. In order to maintain commercial activity we therefore need to continue to exert force in the direction of new business. Since, as we have earlier noted, the degree of security becomes less and less certain as we project ourselves further into the future, we feel the need to protect our assets against adverse contingencies. We do so by levying surcharges on the prices of our products and services to compensate for our risks. If such adverse contingencies do not arise to the extent that they offset the full amount of such levies, then we pocket the difference and call it profit. If we wish to expand our enterprise we invite so-called venture capital to invest or plough part of our profits back into the enterprise thereby exposing ourselves to prospects of greater benefits and greater risks. A correspondingly larger portion of profit will then require to be held against additional contingencies and risk, the balance being expendable and payable as rewards or dividends to those who have invested in the enterprise.

The echo of Geoffrey Chaucer can still be clearly heard: 'Nothing ventured, nothing gained'!

156 Socialism: Cooperative Community Life

Aggregation

What we observe in Nature is the tendency of sub-atomic particles, atoms, molecules, cells, simple and compound organisms and groups of organisms to combine. The above hierarchy is not merely of scale, but chronological. Each level of organization necessarily precedes the larger, more complex one. It is through aggregation that the small gain stature and relative influence, becoming collectively larger and more capable.

Socialization

We recall that the effect of temperature was to facilitate the socialization of substances. The effect of such circulation is to change the nature of relationships. The greater the opportunity for change in relations, the greater the prospects for interaction. Proximity permits interactions that are denied by separation. Interactions imply that two or more forces combine their operations, or cooperate, to cause a single effect. Interactions may be negative, neutral or positive in effect. Negation implies detriment and cost. By this we mean that the forces are competitive and work to offset or negate each other's natural capacities at the cost of loss of power and direction. Neutrality resolves interaction without bias, while positive interactions consolidate power, applying it in a common direction. So the obvious prospect that flows from aggregation is proximate co-occupation of place and time through which contact, affinities, transference and transformation may ensue.

Specialization

The fundamental nature of a relationship is of mutuality. Each party depends upon another for the continuity of the relationship. Dependency becomes established through sharing the attributes of each party to the benefit of others. To the extent that the aggregation is more powerful and diverse than any of its parts, interdependence is a more effective means by which to nurture and protect the integrity of the parts than independence. Each part contributes to the whole according to its native abilities while retaining its own identity to a large degree. That contribution becomes its special function. The division of energy and matter forms the crux of specialized contributions.

Maintenance is advanced through reciprocal services in the form of replacement, substitution or alteration. Thus the expectation of being becomes commuted or extended in time, whether the entity is organic or not. When being becomes tenuous, the replacement, substitution or alteration may advance the entity to a state of transformation whereby something becomes something else.

The individual organism is a self-helping organization of related parts earlier described as satisfying six criteria for entry into the class of the living. Self-sufficiency is the only mode by which solitary organisms may operate. Unless or until they derive greater benefits through association, their potential for evolving into higher, more complex organisms is severely limited.

Mutualism

'All matter has native proclivities for establishing affinities, whether they be chemical, physical or biological', to quote from our discussion of 'life'. While such affinities are not necessarily 'beneficial' in any sense that we can recognize, they all contribute to changes in relations and conditions, many of which enhance the abilities of the 'partners' to do what they do to greater effect or to change the nature of their 'doing'. Where this occurs between organisms it is known as symbiosis and finds expression in cohabitation, scavenging for superfluous food, sharing shelter to the benefit of one or both partners or through parasitism whereby the dependent 'guest' benefits and the host may or may not sustain detrimental effects. Then there is what we might call unconsciously programmed communion whereby like organisms habitually congregate for protection from cold or predators. Many terms are used to denote the instinctive grouping behaviour of animals by species and classes of species.

The collective 'organism' is a mutually supportive ordered progression of independent organisms. The coral is representative of this class of mutualism. Each organism is a discrete individual, a polyp, which cannot survive independent of its colony. In excreting unusable calcium carbonate from its seawater intake and depositing it at its base and around its tube-like body, the polyp builds a protective shield for itself and a foundation for successive generations of polyps.

Appearances of organic growth in inorganic matter are attributable to the mummification of the polyp's organic form. The question as to whether the collective (as a coral reef or any community of organisms) is itself an organism is open to debate.

Lastly, there is the wilful election to socialize exhibited by man for the same reasons but by which means relationships become increasingly more sophisticated and goal-oriented, of which the division of contributions of labour and materials is a conspicuous example. The domestication of crops and animals and the exploitation by man of life-generated substances may be considered to be natural forms of cooperation comparable to the symbiotic relation between ants and aphids, whose honeydew ants harvest in return for protection. Indeed, the exchange of oxygen and carbon dioxide between photosynthesizers and respirers may be viewed in the broadest sense as mutual dependency absent the symbiotic one-to-one relation and with a broadening of the notion of living together. If we now allow the 'butterfly effect' to represent that everything is influenced by every other thing, we must recognize that the loss of any species must be construed to be contributory to some degree to the detriment of all surviving species. What we have been discussing are concepts that are themselves interdependent. We have drawn artificial boundaries around them to clarify for our own purposes phenomena that are essentially interdependent and therefore arguably integral – as living cells are to larger organisms.

Mutation

Mutation, a means by which organisms change their constitutions in response to changes in their environment, affords prospective paths to survival or development that previous forms have been denied. By constitution we are referring to an organism's predisposition to develop and behave in certain ways and not in others. Such programmes are functions of their inherent inventories of parts and relations. Inheritance is the biochemical passage of such inventories from one generation to another through genes. To illustrate this event let us flash back to our understanding of sequence as it applies to numbers. Two governing principles control inheritance: for sequences where the values vary but the operation is unchanging, 'there appears to be a preordination, fate or destiny governing the outcomes of such equations, independent of sequence'; and 'significance in outcome attaches to sequence of operations only where multiplicity of operations occurs'.

Mutation occurs when the content or pattern of inheritance is changed due to losses or transpositions caused by abnormal radiation, chemical processes or temperature changes. Such changes deform or misinform the 'spelling' of the genetic instruction 'duplicate' to read d-en-i-gr-ate or com-plicate.

Changes in inherent parts or patterns open and close opportunities for an organism. We trace such events in order to describe the historical lines of development that we call evolutionary.

Entropy dictates that new developments tend to be more complex than their forebears were, and probability determines that more sophisticated organisms are more likely to survive the effects of novel environmental impacts than specimens that have reverted back to simpler, earlier forms. Thus an organism's abilities to survive and thrive in increasingly more complex environments increases over time, specifically from generation to generation. In so doing its society becomes progressively more complex too.

Predation

Fish do not swim in shoals, or wildebeest migrate in herds because they have a better chance of survival; they have a better chance of survival because they so congregate en masse. That is a function of probability confirmed by statistics. We tend to encounter the survivors and they tend to congregate. In the world of numbers predominance pays dividends. In the world of predation and as a general rule, predators predominate over prey by 'force', while prey predominates over predators in 'numbers'. While predation is the principle threat to animal security, these distinctly differing advantages assure the survival of the hunter and the hunted species.

Premeditation

Man applies the mathematics of probability to his own advantage in times of war when he elects to sail in fleets, fly in squadrons and march in regiments. With such knowledge man can elect to cause events and outcomes that would not otherwise occur.

Through the conscious exercise of will, man hones his skills to produce effects that best serve his interests. Inasmuch as man is both singular and plural, he has difficulties determining where his best interests lie, but the many endure the one, much as they do in the animal world. Where man in the singular is endangered, man in the plural will generally come to his aid. Where man in the plural is in danger, man in the singular will respond. This behaviour, often involving considerable risk, while generally premeditated, may be impulsive. Mutuality of interest and a deeply rooted sense of obligation on the one hand and fear on the other hand confirm that the cooperative spirit in man has not yet risen to the status of an automatic response. Because he is a habituated premeditator and decision-maker, he is unlikely to achieve the status of an automaton in the foreseeable future.

Socialism

Independence is dependent upon self-sufficiency. Independence is founded upon circumstances in which no external collective support is accessible. The basic tenet underlying such an existence is resourcefulness. If an entity has the resources to do something, it may. If not, it will not. So the question arises: under what circumstances would something accomplish an action that it cannot do of its own accord? The straight answer would be if additional resources became available. The fact of plurality provides such resources, not simply in additional numbers but in multiple variety. Recognition of the fact of pluralism necessitates taking this fact into account in the organization of society, or does it?

Ants, bees and termites stand at the pinnacle of socialist achievements. The bee epitomizes Descartes' mechanistic world, a world in which one fertile queen produces female workers and male drones on an as-needed basis to maintain the operation of the machine of state. Bee-haviour is 'clonic'. Each worker bee is identical and programmed identically. By locking out diversity the bee has completely surmounted the social difficulties posed by the individual. With the exception of the queen there are no individuals. All are clones.

Societies are dynamic processes whereby they selectively and collectively capitalize on their assets to overcome adversities. In the absence of compound problems, there is no need for compound (social) solutions. But that is not in the nature of nature. The fact that there are always adversities that cannot be overcome by small numbers drives individuals to become one of a few, and the few to become some of many.

Socialism as an ideal is not so much a state of control as a state of commitment whereby individuals accord a high degree of importance to the welfare and interests of others, because they see that, on balance, it is in their interests to do so! In theory socialism gives citizens a share in possession and the responsibility for decisions that affect their lives, offering a viable alternative path to security to that afforded by default: the necessity to struggle alone for personal survival.

157 Property: Rights to Possession

As is so often the case, we readily recognize antitheses as polar opposite experiences. But between extremities we often have difficulties determining a fitting condition relative to those extremities. Such is the case where the subject of property rights arises.

It is normal to assert claims to ownership of one's body and mind insofar as they cannot be separated from the id that resides in and drives them. Conversely, it takes a wild stretch of the imagination to assert that the universe is the property of a single individual!

However, lesser, indeterminate claims to rights of possession range by degree between the intimately personal and the extravagantly multifarious. Many of our difficulties arise from a deeply rooted bias in favour of ourselves as centres of the universe; the point from which all other points are viewed and from which we establish our presumptions of relative importance.

Societies comprise multiple selves, each of which compete for priority in entitlements to limited resources. Limitations in resources put us at variance with each other in what might otherwise be viewed as an idyllic Garden of Eden: earth. So the need arises to establish a system of priorities in rights of possession in accordance with a generally acceptable scheme of values in the absence of which we can anticipate certain grounds for conflict.

Might is Right

In the animal kingdom the stronger the individual, the stronger his claim upon resources. But even in this milieu there are natural limits; appetites are finite and excessive demands wasteful in material, energy and time! So individual needs govern deeds.

Sleight is Right

Mankind has broken the needs barrier by hoarding excess 'for a rainy day', or to exchange with others for other goods or services. The more he hoards the less there is available to others and the more he can expect to extort in exchange. Thus, we readily recognize denial of any concept of 'a generally acceptable scheme of values' as the basis for our current practices of apportionment!

Presumptions of Entitlement

Inasmuch as plants, animals and mankind enter their respective life cycles, and insofar as the average life span of any species is measurable and can be represented as a normal expectation, then, in an ideal world, all should have a fair chance of survival for their allotted terms. What constitute minimal conditions for survival and a normal life span may be arguable, particularly where critical resources are finite and where populations increase in number; but as a general principle, it is not only reasonable, but desirable, to enable life to abound in quantity and quality for no better reason than that each may be of potential service to others in special and peculiar ways.

Rationale

Distribution of vital resources by ratio, rationing or in accordance with some other rationale is what goes on automatically in a state of scarcity. However, it is only mankind that consciously concerns himself with rational process, and then only when he rises above his native animal proclivities. So apportionment is of interest to man for no better reason than that, in its absence, he reverts to the less agreeable tactics of exercising might or sleight!

Emergence

Man has slowly awakened to heed the need for a strategy for sharing. This interest has generally been cast in forms that engender a strong bias in favour of the individual at the expense of the many. It is as if the doctrines of might and sleight have, very slowly and very reluctantly, been diluted in response to the demands of greater numbers; reluctance arising from the conventional reality that authority was retained and exercised by small minorities.

Man-age-ment

Graduation to manhood, with all that is implied by conscious mentality, requires a formal mechanism for apportionment such that all are, in a manner of speaking, provided for. Ideally no individual or group should be so deprived as to cause it to rise in indignation and assert claims to rights of possession through the operation of might or sleight.

Any such mechanism necessarily needs to recognize claims to entitlements according to a hierarchical system of natural priorities. The rights of the individual to life and sustenance stand paramount. The shared needs of families for shelter are fundamental. The needs of communities for economic opportunities are indisputable. Affordable access to education and healthcare for all is clearly evident. Ultimately the circle of intimate interdependence is closed by assuring that communities of all scales are secure from oppression.

As with any circle it is open to question where one enters the ring and where one places priorities as an individual, but the issue of necessity for maintaining the integrity of the circle is not open to question. To these ends contingency provisions should be fail-safe.

158 Giving: The Foundation of Material and Spiritual Life

If our first impulses in life are selfish we may be forgiven. It is not easy to be delivered into an alien environment without our consent.

I Want!

All we want, after all, is everything! That is what we have been accustomed to all our prenatal lives: nourishment, security, love, comfort, etc.; and it always came from somebody else. How could we not want more of the same, sated as we were in the 'good life'? It is only when we discover that we cannot gain a new 'want' from that somebody that we are confronted by the idea that maybe we have to take to get. So we take. So far so good!

You Want?

But if we take that which others also want, and if it is not possible for all to have according to their needs or wants, we are challenged by the necessity to prevail or make concessions.

We Want

Indeed, we soon discover that most of us want much the same from life, and that life with all its abundant resources just hasn't enough to satisfy our unlimited appetites for more of everything.

We Exchange

We discover competition and are confronted by an unfamiliar and uncomfortable reality: the need to reconcile our differences and compromise in preference to confrontation in conflict. We come to understand priorities whereby we do not need all things concurrently at all times. We learn that in order to gain something we must concede something else.

The Greedy

Some, however, never mature to the point of accepting this principle of reciprocity. They learn how to get and forgo how to give. They account on the accrual basis whereby the more they take the more they draw confidence and believe in their abilities to take more. Success for them is measured quantitatively. Quantity, they believe, gives them access to unlimited quality in life.

Greater Rewards

From the revelation of necessity the idea of sharing takes root as a prescription for reciprocation. Not only do we gain satisfaction, but we gain it in proportion to our giving. We learn a system of accumulating satisfaction that is intrinsically rewarding. A sense of quietude descends upon relationships in which all gain according to their evident and demonstrable needs, sharing equitably proportionate to supplies.

Greater Motivation

Through giving satisfaction transcends that which material gain can offer. We discover that giving has a higher potential for personal fulfilment than material accumulation. In general, saving small indiscretions, most of us measure our successes on the basis of achieving a relative mix of material and spiritual satisfaction.

Greater Happiness

In equating giving with good living we learn to balance our own necessities with those of others. A sense of approximate equilibrium in the relation of every thing to every other thing, most particularly in the relation of every person to every other person whereby no necessity goes unattended, is the only path to personal and social security. Happiness is inseparable from the act of giving.

159 Money: A Facsimile of Opportunity and Power

Money is like language and mathematics; it facilitates transactions between people. Whether the transaction is in the medium of ideas (through language), quantities (through mathematics), or takes an intermediary form, money serves as a common denominator or translatory system of values without which many transactions would be too cumbersome to accomplish.

Money

The intent of money is to provide a neutral means by which to exchange values without the need for correspondence in kind or quantity. A secondary but important consideration is that, being neutral and without bias, money serves to resolve difficulties that would otherwise arise out of the necessity to assess just equivalence in values between disparate goods and services. In this way coercion and fraud are less likely. With the acceptance of a system of money, all we need to do is establish values of all goods and services relative to a socially sanctioned scale of values represented by money. Everyone can then freely establish the price of his own goods and labour in the marketplace. If the products do not sell then assessments of value are too high; if the demand exceeds the supply, then they are too low. With such freedom to trade, social relations are promoted that, all other things being equal, lead to greater social freedoms.

Facsimile

Thus, money becomes a representative currency, a token value system in a common market dealing in commodities that may otherwise have little in common. The marketplace becomes the regulator of economic activity. Speciality markets emerge as foci for trading in goods of like kinds, the fish market, the produce market, etc., which may in turn be aggregated and represented in stock markets as relative fluctuating values that are traded en masse as facsimiles of facsimiles. All that all who participate have to do is to trust in the virtue of virtual values.

Opportunity

Money is liquid opportunity. The rate of exchange of anything for anything else is only a function of compensation in money. Money becomes the key to getting and getting rid of almost anything. The greatest value of accumulated money resides in its release from the need to think about money.

Power

Through the auspices of money one can transpose that which is of relatively little worth into that which is of much greater worth; and by repetition, perhaps with variation, accumulate much that is of great personal value. With much one can accomplish much. With much one can persuade others of the merits of supporting one's interests in expectation of similar gains. Much begets much more. Thus, money is a facsimile for power and opportunity.

Power and opportunity are not ends in themselves but are rather means to further ends. So the open question for all persons who aspire to acquire money always holds: to what ends?

Ends

We have referred to security as being fundamental, but fundamental to what? To continued existence, to wellbeing and to peace of mind? Yes, but peace of mind is hardly a lofty end of itself, but rather a platform from which to pursue higher goals. What are such higher goals but those that lead one away from egocentric pursuits and that enable one to help others to reach out in pursuit of their own goals. Living as giving in support of the needs of others is the ultimate high way. This is not achieved or achievable in the singular, it is relational. To the extent that money facilitates this process as a lubricant, money is put to good purposes.

Insofar as money has become the principle medium through which all else is acquired, the acquisition of money has become an end in itself rather than a means to ends by which to live the good life in modern society. Let us take a look at how that development has impacted our humanity from the standpoint of those who view money primarily as a social exchange currency, as it was introduced to be, rather than a power base that it has largely become. The bottom line with respect to money is not so much how much one has, as what one does with what one has in the context of society. The same maxim applies to many actions taken within a society.

If money is expended (either by society represented by government or by special interests represented by individuals and groups of individuals) in ways that have no detrimental effects upon the interests of others, nobody seems too concerned. The social debate revolves around differences as to where and how distinctions should be drawn between the general interests of society and special interests within it, from which flow presumptions of entitlement to the proceeds of money at work.

160 Capitalism: Competitive Community Life

The United States Declaration of Independence of 1776 and the Constitution that followed in 1787 were utopian socialist manifestos. They were utopian because they sought to promote a form of self-government more closely in accordance with ideals of perfection; and they were socialist because they spoke of the unanimity of 'We the people' and were replete with condemnations of abuse of privilege. There was no hint of capitalism in these documents.

The principle of equality embodied in these manifestos was a radical departure from practical political precedence, bearing as it did upon all with respect to 'certain unalienable rights' including 'life, liberty and the pursuit of happiness' irrespective of the social status of the one, the few and the many in relation to their contemporaries.

What is Capital?

When we speak of capitalism we assume that we all understand what we are talking about. But what is capital? A dictionary would tell us something like 'fit to use in initial position', hence the multiple uses of the term to describe the first letter of a sentence, the first city of a nation, the head as the initiator of human action, etc., all signifying the holding of a predominant position relative to a concomitant body. When we ask ourselves what makes societies work, we are led to the common understanding that it is common interest. As we have already established, interest in a multi-headed society is primarily in security. Security is afforded through the leverage of force, the common denominator of social leverage hinging on economics. We agree upon a common currency for the purposes of exchanging economic favours, and that currency is money.

Where Does Money Come From?

The most accessible sources of money are returns in trade for goods and services rendered. But there is a limit to how much work one man can perform in a day, and that limit exhausts his ability to gain greater wealth. We might characterize this form of wealth development as 'earned' income.

For those who are able to earn more than they need to sustain their business of living, what can they do to supplement their wealth development and consequently their security? The traditional option has been to lend money to those who committed to repaying the debt with an additional dividend in return for the use of the lender's money. That additional dividend was the 'interest' of the lender, which came to be called just that. What were the options open to him to secure against the risk that he exposed his money to when he lent it to others? He had to have some form of leverage over the borrower to cover the eventuality that the borrower failed to repay his debt. What could be more secure than a claim upon the life of the borrower? Coercion became the means by which the lender controlled his investment. The next step to the greater advantage of the lender was to apply coercive pressure upon others without lending them money. In return for the promise of food and shelter, others committed to work. If their earnings were held low enough, the unfortunates would never have the opportunity to be more fortunate; they would be held in bondage until they repaid their debts, a date with destiny that was forever held beyond their means to reach. Slavery was very profitable for those who could afford to lever their security in the form of 'unearned' income. At a certain point they had no need for 'earned' income. Henceforth, money was to come from the labour of other people. This, they maintained, was altogether better than working!

Securing Security

The principle of bondage had become established, but it did not come without a price. How, moneylenders might ask, could they generate a comparable flow of unearned income without the necessity to maintain an army of servants? The answer to that question was as unexpected as it was profitable: lend to those that already have money and who are looking for means by which to do just what the moneylender was doing. What was required, and was initiated by landowners, was a system of entitlement to land that was representative of wealth yet, negotiable like money without need to forfeit the land upon which the wealth was secured. Paper was to replace coinage, and paper was to represent ownership in land and other chattels, property and labour, all of which were negotiable and could be sold to redeem a debt. The formulation of laws, public records and other devices in support of such a principle was to provide the security that secured securities from which position one was free to buy and sell 'securities' in lieu of buying or selling the physical properties that underwrote their value.

Capital Gains

Let us briefly reflect upon the process by which capitalism has become established in the West but failed to take root elsewhere. If a single idea could define the difference between the success and failure of systems to capitalize on their assets it would have to be the institution.

We have spoken of the function that writing serves as evidence to secure against uncertainty in a civilized state. In the West we have extended that interest in security one step further by fostering formats that secure the written word as a function of public policy through the institution of bureaucracy. Public records have been instituted and protected by the people for the benefit of the people. Two other related ideas, civility and service, have served to cement institutional life into the stalwart bastion of Western civilization that it has become. The institution within institutions of the Civil Service system has, to an extraordinary degree, insulated public records from the nefarious influences of political and criminal graft.

Freedom and Servitude

The slaves were freed and a class of quasi-slaves was to take their place. No longer behoven to a single owner, the quasi-slave became bonded in a new way to a new kind of society that, as we shall see, behaved just like the slave owner.

Society laid down in practice that all should be responsible for their own wellbeing. The basis for that practice was the very Constitution that guaranteed 'certain unalienable rights'. These rights were interpreted as the 'principle of equality' in responsibility of each person for securing his own 'life, liberty and the pursuit of happiness'. Competition was to become the modus operandi; a kind of continuous civil war pitting all against all in the pursuit of security through stealth underwritten by wealth.

The Cream Machine

The lifeless never die! So why not form legal entities equivalent to persons but which cannot die? By this device, established bodies of wealth need never expire. All that was needed to be done was to formulate organizational patterns of production from which excess profit could be skimmed on a continuing basis! Into this world was born a 'creature' that could buy, own and sell any kind of security, produce any kind of product and perform any kind of service, be treated by the law as if it were human, yet be 'immortal' and thereby evade estate taxes. The Corporation was incorporated! Henceforth, wealth development was to be detached and held at arm's length from the personal responsibility of 'owners' of the means of production. The primary concern of business enterprise was to afford security for its owners. Only incidentally was it to be concerned with the integrity of its products, processes or services, and then only to the extent that they served the primary purpose.

Other People's Money

Capital is 'kinetic potential' for wealth development, not goods, services and money per se. These latter assets do not earn until they are put to work. Corporations are fuelled by the infusion of money from the multitude of 'quasi-slaves' and 'collectives' that represent special interests.

The exploitation of other people's money (OPM), the opium of capitalism, took on a populist character. Anybody who made a reasonably good living and who had discretionary money to spend was able to 'buy into' the system of accruing 'unearned' income. The 'kinetic potential' represented by such large numbers of wage earners was enormous and the combined assets of the new investors overwhelmed the traditional moneylenders' capacity to control the market forces. A new kind of democratic economy was emerging.

Virtual Wealth

Not only could the ordinary man indulge his fantasy of getting rich quickly; if he played his cards well he could do so without coughing up his own money! He could do unto others what others in high places had been doing unto him for years: get high on OPM!

The stock market was to become the super-market for trading in wealth-equivalence. The populist 'casino' had come to town, every town, and everybody was playing the game 'Capitalist'. Like capital, gains are not 'real' until realized, i.e. cashed-in and used, otherwise all is just 'kinetic potential'. Virtual wealth is a number of things, all of them merely numbers.

Facility

Proficiency in comprehending the dynamic nature of change is not a natural trait. While the forces within the marketplace appear to be independent and so numerous as to make their combined behavioural patterns indecipherable, there is a general understanding that they are somehow interrelated. It is as if the principles embodied in the 'butterfly effect', 'Relativity' and 'Uncertainty' were once more asserting their authorities. Therefore, with respect to the prediction of market activity, investment takes on the character of gambling. Instead of paying the house a dividend, as one would when gambling in a casino, one pays the stockbroker a commission. On balance, like in casino gambling, all would-be investors would be losers if it were not for two contradictory trends that investors rely upon to weight their chances in favour of profitability: growth and inflation. Growth in populations increases demands for goods and services. Investors predicate their 'bets' on the assumption of continuity of growth in demand. Investors bet on the success of an economic system that relies upon growth in productivity. If the system 'wins', they expect to enjoy a share of that success. However, economic success translates to a general increase in the availability of money and credit which, in order to qualify as capital, has to be put to work. The effect of the infusion of new money into the marketplace is to further inflate demands for goods and services beyond that generated by population increase. Thus, inflation may be understood as that process that increases the quantity of money available relative to the quantity of goods and services. This change in the ratio of available money to available goods and services effectively diminishes the purchasing power of money, reducing the margins of profitability in investment. What, one might ask, is the antidote to inflation? The answer is singular: produce more, and faster! Inflation becomes the built-in stimulus to do better next time, and better yet the next, and so on. To survive, the investor must be facile.

Revving Up the Engines of Exchange

On the assumption that the faster you fly, the faster you get there, hastening round trips in the circulation of money becomes key to ensuring that all players have cash flow and cash in hand long enough to make their next investment.

The pressure upon investors to make profits that exceed their losses due to inflation and bad risks is tremendous. While they are now the 'owners' of private enterprise, they cannot afford to be too concerned with the 'whats', 'wheres', 'whys', 'whens' and 'hows' of production. They must focus upon their margins of profit almost exclusively, for these are the only factors that are going to keep them and their enterprise solvent. The unpredictability of the forces of the marketplace dictates a corresponding unpredictability in the behaviour of investors. A fickle economy begets fickle investors. Maximizing returns on investment places an unacceptable strain on growing businesses in need of capital. In order to attract investors, business must appear to offer prospects that are more attractive than offered elsewhere, not merely by direct competitors but by any enterprise that is seeking to tap sources of finance. Conversely, investors are quick to divest themselves of stock in one enterprise when a better prospect presents itself.

Such exposure assures that the growth and profitability of the enterprise cannot be sustained except under the most favourable (lucky or unscrupulous) conditions. Failing businesses become carrion for larger businesses to feed upon, enhancing their capabilities while reducing their competition. 'The higher' the rate of circulation, 'the fewer' in number the survivors.

Pie in the Sky

Commercial airlines cut back on services when competition gets tense. One option commonly adopted is to deny the 'free lunch', a tactic that national airlines seldom employ. Recognizing that one operates on the basis of skimming off unearned profits for otherwise disinterested investors while the other does not, it is not hard to understand who is eating your lunch!

The Sky as a Ceiling

Like the ubiquitous 'chain letter', the prospect of an expanding bubble of automatic wealth development appears to offer unlimited theoretical opportunities for everybody. Surely this is the ultimate economic goal! But as with chain letters, the real gains are to be enjoyed only by those who initiate the process or who buy and sell during the early cycles of circulation. Compounding investments only work for a small minority of entrepreneurs and then only if they are lucky! A ceiling of clouds seems to portend in an otherwise seamless and infinite sky!

Capital Losses

Given the extent to which the capitalist model has been so successful in the West, we should now project ourselves forward into a world in which many more nations aspire to emulate, and may actually out-perform, the West in some important respects. What would such a world look like, what would the prospects for achieving such a scaled-up version of capitalism be and for how long would such a behemoth juggernaut be viable?

A few key indicators tell us that it is simply not going to happen, anyway in the format that capitalism is currently being conducted. When 6 per cent of the world's population consumes 25 per cent of the world's total output of non-renewable energy resources, it doesn't take higher mathematics to project what will happen when demand and consumption rise by a factor of four!

In our discussion of corruption we outlined the gravity of the situation concerning the rampant disregard for sustainable economies and the inevitable consequences that would flow.

The Life Cycle of Capitalism

Once again it is befitting to employ the bell curve, in this case to characterize the life cycle of capitalism. The early effects of applying capitalist principles had the effect of enabling the generation and distribution of wealth and the exercise of authority over a broader base than conventional pyramidal power structures permitted. What enabled the new barons of industry and commerce to prosper was a surfeit of cheap labour and materials. Economic growth was remarkable for the few. As the number of 'capitalists' increased, these surplus resources began to be drawn down to increase production in response to burgeoning consumer demands. Conflicts of interest between production ownership and labour abounded. Was capitalist economics falling short of its promise? Had capitalism peaked and, if so, what was on the dark side of the bell?

A Flaw in the Floor

In discussing security earlier we established that the prospects of change and conflicts of interest rendered diminishing returns of security when projected into the distant future. Further, we established the effect of such uncertainty was to deny the ultimate security: peace of mind!

To project a future for capitalism we need to take stock of our progress and anticipate a time when the resources available today will no longer be available in quantities sufficient to sustain growth. The value of capital and therefore capitalism will still reside in its potential. As potential to deliver the 'goods' diminishes, so too will the viability of a culture predicated upon capitalism. The prospect of structural failure of a system that relies so fundamentally upon the thesis of ever-increasing rates of exploitation, production and consumption is not simply theoretical, but imminent. Under such circumstances, no amount of capital investment will afford security.

At such a time the emphasis of economics will necessarily require to be deflected away from the accumulation of wealth as the measure of security, to the enhancement of wellbeing measured on a different scale. This necessitates a change in focus from means to ends. In modern society the means by which wealth is acquired is well understood to the extent that it has become a goal in itself rather than a means to higher, worthier goals.

Looking Up

What are the prospects for an economic system that cannot sustain itself due to change and conflicts of interest? The optimist will look for positive signs of changes well in advance of a crisis. What economic model could conceivably emerge that would be less dependent upon growth in production and the continued exploitation of diminishing supplies of materials?

The first signs of positive changes may be characterized as engineered of necessity. Materials were to be evaluated for their maximum potential to perform requisite functions without excess.

The concept of accomplishing more through utilizing less was slowly to become the *sine qua non* for the employment of labour and materials. The benefits were to effect a stretching out of the time frame within which these resources would otherwise be utilized.

The second quantum change that emerged was a gradual reduction in the volume of goods produced and a corresponding increase in the level of services rendered, the effects of which were to maintain high levels of employment while consuming fewer resources.

The third beneficial signal was of an increased interest in learning made necessary by the demands of service industries. Education enhanced abilities to perform in the foregoing sectors. Incidentally, and quite unexpectedly, industrialized nations experienced a significant decline in the annual growth rates of populations, thereby further reconciling reductions in demand with the practical difficulties of maintaining long-term growth in productivity.

The gross effect of these net changes suggests a gradual move towards the correlation of demand and supply with an interest in achieving a sustainable economy within available means.

Looking Down

The pessimist will argue that, notwithstanding the above, the ugly side of capitalism continues to exploit all there is for all its worth in the short term. Military adventures against Third World countries on bogus pretexts continue unabated, so much so that it has become hard to distinguish between what capitalist powers do today and what their nemeses, the communists, did in yesteryears. The parallel but unparalleled expedient of unscrupulous greed feeds on! That the greedy and their military minions are motivated by fear does not excuse their violation of the integrity of the living world or their denial of the natural law favouring equilibrium.

Anomaly or Normality?

As is the case where any system becomes established through the vicissitudes of trial and error, capitalism has acquired a degree of sophistication that has enabled the able to compound their benefits exponentially by comparison with the less able. In doing so, not only is this phenomenon cited as further confirmation of the correctness of Alfred Russel Wallace's theory of 'the survival of the fittest', it is pretentiously used to justify further distensions of this trend.

What is conspicuously absent from this discussion is acknowledgement that the evolutionary path that man is embarking upon is distinctly different from that of any other species, giving rise by implication to responsibilities far beyond any imagined by the most forward-looking minds of a generation or two earlier.

Projections of a future for capitalism along the lines that we have discussed strongly suggest a widening gulf between the life expectations and experiences of the fortunate and those of the less fortunate. Such a schism as is being wrought between the two cultures, if allowed to extend indeterminately, could transpire to be interpreted as a clear vindication of the thesis underlying Wallace's other perspicacious postulate: 'the tendency of varieties to depart indefinitely from the original type' – to wit a species split! The open question is whether, through extenuating these developments, mankind would allow his kind to ostensibly effect an irreconcilable division between man and superman.

We shall pick up the subject of master-race theories in our later discussions of religion and the supernatural. Suffice to say there are emergent signs suggesting that evolutionary genetic mutation is already underway that, if allowed to take its natural course, will not only rend civilization as we know it asunder, but will provide for those who stand to benefit from such a course a compelling rationale in support of their interests. That is to say, that theirs is the correct and inevitable path for human development to take, notwithstanding extraordinary expenditures of life and resources in the cause of democracy to deny that this should be our destiny.

After Capitalism?

The goal of capitalism, if one can say that there is one, is to make money govern human action, a concept that is in conflict with more lofty goals of individuals and society. The fallacy of capitalism is that the more people have, the more they want; and we know where most capital comes from!

Removal of the monetary incentive from community life raises men to a state where they are judged by their thoughts and actions rather than by their accumulated wealth and power. The ultimate goal of 'high rolling' is to roll towards higher goals. We should all get used to the idea that special interests may not be working for the general interest and proceed from there.

Why should society be limited to a simple-minded 'either/or' choice with respect to enjoying the benefits of capitalism or socialism when we not only can afford the best of both, we cannot afford not to have them? Individual interests and social necessities persist whether we recognize them culturally or not. Capitalism is the religion of the insecure. The high priests of capitalism do not find their Holy Grail or happiness through the accumulation of wealth. We need to reiterate the question of whose interests are greater, those of the leaders or those of the followers.

The reliance upon objective evidence to guide decision-making is usually required by law as a necessary precaution to ensure accountability for actions taken in the public sector. We acknowledge the tendency for unsophisticated employees and elected or appointed officials to informalize their decision-making through allowing emotions to cloud their judgements. The effect of this informality is to undermine the network of legal controls, checks and balances that have progressively underpinned the evolution of capitalism since the Reformation, the Age of Enlightenment and the industrial revolution. Such laxity degrades the rule of law to a form that is perhaps better described as the lore of illegitimacy with potential for disastrous economic and social consequences.

The reason that capitalism has attained ascendancy over other systems in recent times is that it is not as adamantly committed to the success of individuals or as aggressively antisocial as some would like to paint it. The focus of its energy is directed towards the development of capital and wealth that are as critical to the interests of individuals as to society. Capitalism is not as 'extreme' a disposition as it may appear unless its processes severely disadvantage certain segments of society. The fact that they do, as attested to by great disparities in the distribution of wealth and privilege, is as good a reason for concern for the average individual as it is for society.

Our youth is not short on courage, as exhibited in the extreme sports that it indulges in; it is short on those critical faculties that reflect upon and cause us to act upon the issues of the day that undermine humanity. Those issues that, in the 1960s, prompted righteous indignation in our youth are today relentlessly countered and suppressed by the proponents of the technocratic society who strive with religious zeal to eulogize the doctrine 'might is right'.

Reformation

A revelation as to the natural process of succession to capitalism is pending. What is apparent is that its days as the stalwart bulwark of superior authority are numbered. Any successor system will necessarily need to overcome the vexing challenges of forestalling material and spiritual poverty. Future societies' determinations as to acceptable bounds of social behaviour may well place the aggressive pursuit of wealth for personal gain on the negative side of the balance sheet.

There will come a need to reschedule production on the basis of need, not greed that expands exponentially at the expense of need. Man, in manufacturing quantities on the basis of their saleability rather than in response to demonstrable need, has skewed nature's equation matching force with equal force. You don't see lions hunting after they have just had a good meal!

The word 'communism' has fallen into disrepute as determined by the results of those social experiments that have conspicuously failed to live up to their highest ideals. Similarly, 'socialism' is viewed as that doctrine that has been mismanaged when opportunities for it to flourish arose. Nevertheless, in principle they afford paths to a levelling of opportunities that, if achieved, would negate the great disparities in wealth and power distribution that the alternative systems of 'capitalism' and 'fascism' promote, which by their natures turn men against each other and need into greed.

Forces favouring a sustainable economy and those attempting to maintain an unsustainable one are counterpoised political polemics destined through conflicts of interest to lead mankind where neither camp wants to go. Minimally it appears essential to re-examine and reinvent the fundamental postulates that underpin capitalism – in order to place the highest priorities on human capital.

161 Poverty: Sparsity of Resources

Poverty reveals acquaintance with the low end of economic productivity. Let us examine this aspect of the economic spectrum as it affects mankind.

Material Poverty

Absence of quantity, quality, variety and choice defines the plight of the materially disadvantaged. Confronted by an apparent inability to enhance prevailing conditions generates humiliation, an inauspicious precursor to despair and desperation.

When one stands at the South Pole no direction points to the south. There are no compromise directions to the east or west to consider; the only bearing available is north, an oft-unbearable prospect! To be sure, there are the theoretical options of moving up to an alleged heaven or down to hell, but there are no assurances that these places exist, not to question whether they are accessible to the opportunity-starved. In any event we do not see lines of volunteers to enter either of these domains when people still cling to life on earth, however tenuously. Such is the status of the destitute; at a loss to distinguish the right way when all ways appear to point in the same direction in an environment devoid of prospects of beneficial change. The essential constituents of life flow in a stream of matter. When the stream dries up, life withers and dies.

Material poverty denies mankind humanity. All that we take for granted has not been so for those in greatest need. Such a life endures where material needs remain unattended.

Social Poverty

Abject material poverty denies not only the ability of an organism to sustain itself, but its ability to be of service to others. Material paucity cannot be construed to be in the interest of any living system, or to offer prospects of greater variety and diversity through reproduction, except as young life may be better equipped to suffer privation and thereby maintain a tenuous hold on life pending improved conditions. Reproduction is often the last desperate resort of an organism, as if to exhibit faith in a better life thereafter.

Intellectual Poverty

Knowledge of poverty is tacit acknowledgement of an individual's relation to poverty. The inability to acknowledge such connections reveals intellectual poverty.

As to whether one attempts to exploit, ignore or ameliorate perceived poverty depends upon circumstances and an individual's personal disposition towards others. Intellectual poverty denies resourcefulness. In the absence of intellectual resources no concept of possibility is possible. Denial portends denial.

Spiritual Poverty

Where no human sensitivity to poverty resides, there abides spiritual poverty. Knowledge carries responsibilities. We cannot individually respond to all pleas for relief from poverty, but we can act in ways to reduce its impact upon others by economizing in our own demands and thereby advancing the redistribution of resources. Initiative is the principal qualification for charitable action.

Moral Poverty

Concerning the unwillingness of society to underwrite the basic needs of the indigent, denial places the physically needy in an excruciating moral dilemma: the choice between begging and stealing. Insofar as most cannot secure subsistence from begging, they are led to the only alternative that nature offers them under her laws, in flagrant defiance of laws legitimated by man. When apprehended the needy are criminalized, an act that, according to our social mores, justifies detention and the provision of food, shelter and medical attention. What kind of extraordinary path in logic and cultural depravity brings us to act in this way? The conspicuous neglect of basic human needs by the privileged in any society clearly signifies intellectual, spiritual and moral poverty. Failure to act upon the evidence of dire need through denying relief is a deeply rooted and justifiable cause of resentment and social instability. So the three necessary steps required to relieve poverty appear to be comprehension, compassion and commitment.

We would do well to note the attention that was brought to bear upon the condition of the needy by Charles Dickens, William Wordsworth, Engels and others; a focus that has been largely marginalized or presumed to be mitigated by token social programmes that pay a pittance to relieve the real problems of the world's underclasses.

If one doesn't have a serious interest in life beyond making money and exercising authority, what kind of human being is this if not poverty stricken?

162 Cost: Whatever Must Be Given

Costs are the antithesis of benefits in an equation that always resolves in a relation of zero-gain and zero-loss in the balance sheet of nature. Nature does not gain or lose one atom of matter or one electron volt of energy except as they may be converted from one form to the other. The reader may object that this is not consistent with his experience, voluntary exchanges always being undertaken with the prospect of gain for each party. But, as we shall see, such a premise does not take into account the full extent of the stream of events. What we can construe from the above is that all notions of costs and benefits are concomitant, occurring collaterally as products of circumstance and perception.

Cost in Place

Cost is invariably sustained in place. Whether by natural cause or manoeuvre, changes occur on location the effects of which register locally. Should two similar events occur simultaneously at two locations, the effects would likely differ insofar as they affect their immediate environments. If snow falls in the Appalachian Mountains, it is the local residents who pay the costs of additional fuel to heat their homes, not those in the Sierra Nevada.

Cost in Time

As a general rule, the further downstream in time that one acquires a product from its point of origin, the greater the cost to the consumer. This is because additional costs get surcharged by every hand through which it passes for shipping and handling, stocking, distribution, wholesaling and, at the end of the time line, retailing.

If one builds a house at a certain point in time, one cannot expect that the cost of a second house built twenty years later would be the same as the cost of the earlier house, even when built at the same location. This is because costs reflect the availability of materials, labour and services that undergo change. Costs tend to inflate over time due to more competitive demands that arise from the depletion of material stocks, pay increases and a general increase in the complexity and therefore cost of doing business. Another way of looking at the same phenomenon is to recognize that the value of money tends to deflate over time.

Cost of Circumstances

As we have earlier suggested, all is in a state of flux. Some events occur suddenly, others more slowly, but the effect of such a dynamic is to render reliability of costs equivocal.

Put yourself in the place of a one-dollar bill. How should a dollar place a value upon itself when all events around it are in a state of change? One day the price of one dozen eggs is X, the next day Y, while down the road and around the corner the price is Z! We are back to physics lesson 101! Nothing is static long enough to rely upon with respect to either place or time. Circumstances circumvent certainty! So how can anybody keep track of 'real' costs in such an untrustworthy environment? Well, as it turns out, they cannot. Reality is surreal in the world of costs. The only costs that count are here-and-now costs.

What we inevitably find ourselves doing when talking or writing about costs is to take an indirect or roundabout course in order to reconcile our numbers with uncertainty. The accountant attempts to identify all credits on one side of his ledger and weigh them against all debits on the other side. He elects to over-simplify his accounting whereby he ignores that of which he is unaware or does not wish to include. Do you remember Occam's razor? Whether an accountant is a conscientious professional or a guileful miscreant, he always manages to make his balance sheets balance. His debit column tallies with his credit column only to the extent that his omissions on each side of the ledger balance, at which point he believes that he has justified the account.

He always makes accounts 'work' because that is the 'work' that he always does. Nature accounts for everything, whether included in or omitted from accountants' books.

Accountants cannot resist the temptation to 'fix' some parameter of the firmament in order that they can establish a basis for comparison for their arithmetic equations. We qualify our language with respect to certitude by prefacing our discussions of cost with references to a specific place at a specific time. Nothing is very clear but that we simplify to make it so.

Nature's Balance of Payments

The measure of what costs really are (distinguished from what one pays or accounts for) is represented in the correct answer to the question: what would it take to restore a conversion of matter or energy (for that is what all exchanges are) to its original, neutral (zero-gain, zero-loss) balanced state? When one looks at accounting in this way, we discover hidden costs and losses (negative benefits).

When one looks (for example) at the cost of restoring petroleum reserves to their natural, balanced state, the offset cost of gasoline is astronomically high, perhaps $1 million per gallon! The difference between the sales price of gasoline and the cost of sustaining supplies is a true reflection of the extent to which man is responsible for misrepresenting his accounts. He skews the natural balance in his relation to nature. Unlike the scientist, the accountant attempts to make his numbers correspond to his perceptions (theories) of truth or desire, not to any demonstrable truth regarding the nature of his environment.

Attempting to take into account the full extent of changing circumstances is a more rational, just approach to assessing the cost/benefit equation with respect to man's interference with nature's delicate system of accounting.

Bridging Space-Time Divides

We return to our house-building scenario for a moment. Simplistically, by attaching a unit of actual cost to each element installed in a building and multiplying the cost of each element by the quantities used, we can combine their subtotals to represent the total cost of construction. Such records are of considerable value since they constitute evidence that may be used to refute any improper claim by one party against another, or for establishing tax deductions, etc.

Insofar as many homes are built over time, a statistical composite of all such records is an invaluable reference for all those preparing to estimate the costs of similar undertakings. All that is missing are means by which to adjust the data to take into account differences in geography and chronology. At this point we take a leaf out of the accountant's books and fix the parameters of place and time for each class of like element or project by assigning index numbers to locations and dates.

We can now better navigate the seas of uncertainty through the use of uniform cost data to estimate the probable cost of building any home anywhere, adjusting for location and time by multiplying our cost data by appropriate index numbers.

What we cannot do is put a price on the cost of contingencies that may arise after the date of our estimates but prior to completion of a new structure. The best that we can do is to add money to the construction budget, call it a contingency sum and hope that it will be adequate to cover unforeseeable additional costs that may arise in the near-term future.

Dark Matter and Hidden Agendas

To the extent that we include or omit to take into consideration significant elements of an equation, we dispense with accuracy.

So far, the costs of our model have been relatively easy to quantify since to a large degree we have been dealing with tangible things fixed in space and time. All of the above has been predicated upon assumptions that people act diligently and honourably, but what if they do not?

The implications here are far-reaching. If we accept that all equations should equate, then we need to look for equitable conditions in all transactions. If we find them we have found a certain manifestation of truth and justice. If we do not find them, or find certain inequities in a situation, then we need to search more diligently and account for the underlying causes.

Let us take a quick look at so-called 'free trade' to enquire what about it is truly free. If we identify the cost components built into products manufactured in highly industrialized countries we can characterize the costs of materials, labour, insurance, transportation and taxes as being 'high', 'high', 'high', 'high' and 'high' respectively. If we then compare similar products manufactured in less-developed countries we can characterize the same cost components as being relatively 'low', 'low', 'low', 'low' and 'low'.

If identical automobiles are built in North America and Asia, when they pass in the middle of the Pacific in transit from one to the other they are 'intrinsically' identical in value. If they are intrinsically identical in value in mid-Pacific, how can it be advantageous to ship them across the ocean? Well, it is and it isn't! It is advantageous to ship the Asian models to America because they have 'extrinsic value' added, or from the American point of view, their costs are discounted while the opposite is the case with respect to shipping the American models in the opposite direction. The latter models have 'extrinsic costs' added that make them non-competitive in the Asian market. A measure of exploitation is necessary to justify exportation. But who is exploiting whom, and how?

The effect of a disproportionate distribution of wealth is for the benefits in goods and services to flow from weak economies to strong economies, and for money to flow from strong to weak economies. Dollars roll downhill! Less-developed countries gain strength from the infusion of money into their economies from which they derive higher standards of living. They indulge in their newfound prosperity by which means they reduce their competitive advantages over more highly industrialized countries.

Industrialized countries profit from the lower production costs of foreign goods. One effect is to free capital for investment in new ventures; technologies that are unavailable in the less-developed countries because of lack of capital and knowledge. By such means the industrialized countries create *ipso facto* monopolies in these patent-protected technologies that are designed to lever ever-greater economic advantages over the disadvantaged.

Because the American automobile doesn't sell well in Asia the Americans withdraw this item from their export inventories and substitute a line of products, say passenger aeroplanes that are not manufactured in Asia, for which they can charge exorbitant prices. Excess profits from such sales get ploughed back into manufacturing the next generation of technological wizardry.

Equitable Interests

If we fail to pay our taxes from our gross earnings we may expect detrimental consequences. What we are alluding to is acknowledgement that our gross earnings are not all 'ours'. Only our net earnings after the deduction of costs and taxes are truly 'ours'. Taxes are membership dues payable to society in proportion to benefits received in accordance with the social contract.

Inasmuch as we have recognized that the social contract conveys privileges and obligations that apply uniformly to the citizenry, such uniformity can be interpreted to convey an equitable interest in the structure of society. This is acknowledged in the one-person-one-vote and the equal protection principles. If equal rights pertain to structure one may argue should they not also apply to substance? If to substance, do they not logically dictate equity in opportunities, including equal access to the supply and distribution of goods?

Regarding the issue of equity, there are two contentions concerning the probity of 'free trade'. The first maintains that unrestricted trade brings benefits to all parties and that such a system tends to operate to the effect of improving economic conditions in all segments of society. The second maintains that there is need for a correctional device in pricing structures whereby weaker economies would receive additional compensation for their products to offset the loss of benefits enjoyed in stronger economies were they to make the same products. From such a strategy it is assumed that the disparity between the rich and poor economic systems would be gradually reduced and eventually dissolve. There are a couple of thorny problems associated with these premises. In the first case the rate of economic gain is much greater for the rich than for the poor, thus such a policy increases the disparity in means between the two, notwithstanding efforts applied to the contrary. In the latter view there is no assurance that any added cost increment would be passed down to the labouring classes as compensation in lieu of benefits not received. On a more sombre note, if these strategies were to be played out to their logical conclusions, either the poor could not afford to buy anything from the rich, or, in the event of the establishment of artificially uniform costs based on what we have called 'intrinsic values', there would be little incentive to trade anything to anyone! It is the labouring class that pays the concealed costs, directly or indirectly, while the privileged class makes off with the booty! Unpaid dues are retained in memory and build up resentment until they reach critical proportions, at which time critical actions are taken to correct imbalance.

Stop-Gap Equations

On a pragmatic level what happens therefore is that trade is conducted on unequal, quasi-coercive terms until exploitation becomes so egregious that one party imposes tariffs, economic sanctions or trade embargoes against another in order to enforce more favourable terms. Sanctions are an effective way of saying 'we do not accept the terms of your proposition'. While this bargaining strategy may work in the short-term, it cannot be sustained indefinitely by a weak economy and becomes self-injurious if extended for a long duration.

Imbalance eventually leads to deficit spending whereby one party becomes so indebted to another that it is unable to meet its obligations and defaults. Something is awry with regard to the formation of such inequitable relations.

Ideal Equations

The goal of idealists is to be intrinsically fair. To account on a truly equitable basis it is therefore necessary to negate the influence of all 'extrinsic' values. It is not coincidental that the great preponderance of trade is transacted between prosperous states where opportunities for inequitable leverage are minimal. In order to maximize the volume of trade between states the ideal would be for each to have equal means with which to buy the products of others.

In answer to our earlier question, 'what should "free trade" be free of?', truly free trade is free of inequities.

Expedient Remedy

What is required to obviate such distorted relations is a constitutionally protected correctional bias whereby greater benefits flow to the disadvantaged than to the wealthy, the intent of which is to gradually reduce disparities and attendant animosities. Accomplishment of such ends is a function of civilized behaviour not wealth-propelled power politics. It requires a modicum of maturity to accept that the long-term interests of all people depend upon their enjoying approximate equality in status: opportunities, rights and prospects. The means by which such a balanced corrective mechanism could be implemented might be through the introduction of a form of indexing that takes into account relative places and times. The degree of bias or rate of adjustment would need to take into account the needs for all parties to adjust to the idea of justice. Such a means of accounting is ultimately governed by political action.

The obvious means to that end is a graduated system of taxation that works for the general good while advancing the standard of living of the poor. To that end tax loopholes for the rich would need to be closed. It is inconceivable that this could be accomplished under a system where legislators require enormous sums of money to get elected, and spend much of their time in elective office raising funds from industrial and commercial powerhouses!

In the United States it has been characterized that a Republican is a Democrat without a conscience, while a Democrat is a Republican without initiative. The initiative rests with the majority of the electorate that constitutes the bulk of the consumer market: rich in potential for voting for a more equitable system to become a political reality.

So it is in the rest of the world.

163 Depreciation: That Which is Lost

The Allegory of Atlantis

The state of Atlantis, as is the custom in democracies, collects taxes and applies the proceeds as it deems most beneficial to the citizens of Atlantis. This process allocates money for the construction of school facilities. Under certain conditions the money is transferred to local school districts to be disbursed on a pro-rata basis against the value of goods and services received as the design and construction work proceeds.

Orange Blossom School District, in justifying its needs for a new elementary school to the state, qualifies to receive construction funding and hires an architect utilizing a non-negotiable form of agreement promulgated by the state, seen to be in its best interests, and recognized by the architect to be prejudicial and onerous.

The architect, Leonardo, selected by Orange on the basis of experience and reputation, signs the agreement to provide professional services in exchange for a fee established as an agreed percentage of the cost of construction. He receives Orange's instructions as to its educational and physical requirements, designs the new school buildings and estimates the probable cost of construction, all to the full satisfaction of Orange.

Several years elapse before funds are made available to Orange by Atlantis, during which time the personnel involved in the project at Orange retire or have moved on. When funding is finally authorized, inexperienced people are appointed to administer the project for Orange and Orange invites competitive bids for construction.

Now let us imagine that the economic climate for business opportunities is such that work is scarce. In order to win a contract a builder knows that he must outbid hungry competitors. Perhaps the viability of his or another firm is at stake. He has real estate and equipment mortgage payments, insurance premiums, office payroll and other overhead costs to pay. He needs this work to stay in business. He is willing to take greater risks than would be necessary in a more buoyant economy. To survive in the construction marketplace he must be the successful bidder and secure the contract. Then he will have cash flow. Then he can reassess his position, work to improve his efficiency, run the project in his best financial interests, and eat. What are his options? During the bidding phase he can diligently estimate the cost of doing the work and tender his bid on that basis; he can further reduce his bid by reducing or omitting his profit margin, bidding at cost to him; or he can bid below market costs, understanding that he must somehow retrieve these losses after the award. These options are progressively more risky.

The documents prepared by Leonardo spell out specific materials and products defining their quantities on the drawings, and quality and performance attributes in a document called a construction specification. The process of inviting bids offers opportunities to competing licensed builders to seek clarification of the contract documents and to offer substitute products to those specified, subject to Leonardo's written response transmitted to all qualified bidders.

Bids are received and evaluated by Leonardo and Orange, and Orange enters a contract with the successful bidder. Under state law a contract must be awarded to the lowest bona fide bidder.

Able Bodied Contractors has competed for this contract successfully. They bind themselves to the contract conditions while recognizing that there are innumerable uncertainties as to the sequence, timing and real costs of performance. The contractor is a bandleader not an instrumentalist, yet is committed to complete work and to guarantee the performance of others. He is at risk. He is also in business to stay in business, which necessitates making a reasonable profit on each project undertaken.

Under normal circumstances such an agreement is implemented with reasonable diligence by all parties in good faith. End of story. In the event of the attendant harsh bidding conditions, let us say that Able has taken the aforesaid risks in order to procure the work.

The construction phase commences. Leonardo's prime responsibility in administering the construction contract is to verify that Orange gets full value for its investment while being fair to both parties. As a means to this end the architect requires that the contractor submit for Leonardo's approval and prior to incorporation in the work, verification of the specific products proposed for installation in the project. The contractor is now under great pressure to recoup his paper losses and he likely sees this process as his path to his financial salvation.

There are two mechanisms commonly used to change the terms of a construction contract; the first involves no change in cost or time of performance, the second involves such changes. Both offer windows of opportunity for the contractor as follows.

In making his submittals to Leonardo, Able maintains that he bid products other than those specified but having the same serviceable characteristics. He seeks approval of these products contending their functional equivalence to those specified, often citing procurement difficulties as a pretext. There is no discussion of changes of cost or time in the contract, nor does he seek additional payment. Such submittals are generally found not to conform to the performance criteria that were established by the architect and invariably cost the contractor less to purchase than the items specified. This is the silent conduit to economic recovery. The burden is ostensibly on the contractor to satisfy the architect of the merits of his substitutions but there is no obligation on the part of the architect to review, far less accept such substitutions. Leonardo and Orange are wary of such procedures, if used, but they are common ploys by contractors under such circumstances.

By inundating Leonardo with written requests for clarification Able seeks occasion to reveal uncertainties that could lead him to claims for additional monies, which claims he will promptly prosecute. Pursuant to this process Leonardo is confronted by an inordinate amount of paperwork, the great bulk of which is trivial or inconsequential. Nevertheless he responds promptly and conscientiously in writing. The Owner, receiving copies of all project correspondence, is likely to become concerned about the extraordinary paper flow, believing that something might be going awry. It is indeed.

For the unsophisticated Owner the processes of construction can often seem baffling and frustrating. A contractor in the situation described above is relying upon this. Let us suppose for a moment that Leonardo, being a dedicated professional, has not shown interest in the political relations of parties, and that the ear of the unsophisticated Owner has newfound sympathy with the frequent complaints of Able. Any suggestion by Able that Leonardo may be at fault will likely also find sympathy in the ear of the unsophisticated Owner since this will serve to explain for him the extraordinary paper flow. The Owner and contractor now feel comfortable discussing the project in the absence of Leonardo notwithstanding provisions in their contracts to communicate only through the architect. The contractor suggests to Orange that under the circumstances his interests may be well served by obtaining a second opinion, that of a 'disinterested' architect. After due consideration and discussion with its lawyer, and perhaps others, Orange retains architect No.2, serving notice upon Leonardo that it is doing so but that the two architects are not to communicate on the matter. The 'disinterested' architect is to review all correspondence and to sit in on all subsequent meetings but will not participate directly. It is significant that architect No.2 elects not to openly disclose a limit to his interests to that of providing secondary opinion, a disclosure that any honourable professional would normally make without question.

Leonardo protests but acquiesces in the knowledge that no serious flaws in his work had been substantiated and that he is confident that he will be able to continue to perform his responsibilities honourably and with distinction under his agreement with Orange. He notifies his insurance carriers of the development.

The 'disinterested' architect now commences his work for compensation, participating in meetings involving lawyers, insurance carriers and all the principle parties at interest to the project except, and unbeknownst to, Leonardo. Shortly thereafter Leonardo receives a registered letter from Orange's lawyers terminating his agreement with Orange alleging grounds which Leonardo knows clearly to be unfounded and in breach of specific conditions for termination in their agreement.

Leonardo notifies his insurance carriers and retains a lawyer, expending more money than the balance of fees due him under his agreement with Orange. No beneficial progress is evident. He caps expenditures. The 'disinterested' architect is now retained by Orange to substitute for the absent Leonardo. Since he is clearly no longer 'disinterested' we shall call him Machiavelli.

We shall interrupt the story for a moment to take stock of what has transpired. Construction has proceeded without major technical difficulties, is 25 per cent completed and no change orders have been issued to change the scope, cost or time of completion of the project. Minor delays have arisen due to disputes involving the neighbourhood, utility companies and Orange concerning payment for improvements to a road serving the new school but these are not attributable to the architect or the contractor. What has transpired of significance are changes in the relations of the parties, and the nature of extra-contractual decisions that have been made. Let us continue.

Orange has made an irreversible decision, to change architects. In so doing it has created an additional and momentous interest, the force of which is yet to be revealed: the vindication of that decision. The actors at Orange feel compelled to prepare for the eventuality of an accounting for their action, on political, person, legal and economic grounds.

For Able the dark cloud of Leonardo's strict interpretation of Able's contract with Orange has blown away. He finds in Machiavelli a more sympathetic surrogate; after all they are both simply trying to make a living. The door has opened for Able to pursue his hidden agenda, the compound addition of costs to the scope of the work.

Machiavelli's machinations have borne fruit. The fruit is for the picking. He is first and foremost a businessman. It is incumbent upon him to maintain the confidence of Orange that Leonardo let slip away. To this end he sets out to justify for Orange the wisdom of its decision.

There follows a stream of demands from Able for changes that will, he posits, benefit the Owner and expedite the work. Cost adjustments are proposed, some reducing the scope of the work and translating to a monetary credit to the Owner, others 'regrettably' adding costs but 'necessitated' by the alleged 'poor performance' of the now absent Leonardo. Additions and deductions are much the same for Able for he gives less than he takes in either event, inflating costs and depreciating credits as each case may suggest.

Machiavelli's responsibilities as the surrogate architect include the review and evaluation of Able's demands. The accumulation of changes to the construction contract work to underpin Orange's decision to hire Machiavelli while keeping Able's morale high. Reviews are cursory and Machiavelli's written endorsements are forwarded to Orange for approval.

Orange, observing that relations between the Owner, architect and contractor are repaired, and that construction continues, authorizes changes, not knowing where additional funds are coming from, hoping that they will come from Atlantis but entertaining the contingency that they may wish to try to extract them from Leonardo's insurance carrier, which, coincidentally, is the same that provides insurance coverage to Machiavelli.

All the while Atlantis is three steps behind. Requests for the approval and funding of additive changes appear but Atlantis has not been informed of a more fundamental change: that of the architects.

Building continues with the substitution of materials, products and systems of inferior quality and performance. The design integrity of the whole is not recognizable to the players as they piecemeal their ways to the fulfilment of their respective goals, and its abrogation is therefore not recognizable either. The damage is done.

Unsophisticated Orange still doesn't appear to know what has been going on.

The construction is completed and Orange assesses the cost over-runs of the project. By Orange's own accounting it has committed itself to additional expenditures in excess of 25 per cent of the original construction cost of the project. Orange has exchanged its small outstanding fee obligation to Leonardo for a debt of millions of dollars and the meter is still ticking with pending litigation, like a time bomb!

The story is played out with claims and counter-claims, and the revelation of an extraordinary complexity of ignorance, incompetence, arrogance, intrigue, collusion, fraud, conspiracy, conflicts of interest and legal shenanigans, but, for the purposes of our discussion, casts no new light on the nature of the relation between costs and benefits.

The costs of the project have accumulated in the form of additive costs of new work, justified or not as the case may be, that include surcharges by the contractor above the market costs of performing the work and costs attributable to less than market value credits given the Owner for deductible cost changes. In addition there are compensation costs for the second opinion, and costs of Machiavelli's subsequent services. Additionally there are costs attributable to loss of value in the integrity of the finished work; legal costs accrued to date by all parties; the costs of additional insurance; and as yet unknown, open-ended but potentially substantial costs of resolving disputes through legal process.

Value, which is the amount of benefit resulting from an event, may be assumed to be approximately equal to costs, but let us look to whom these benefits have been disbursed rather than being applied to the project: Able has fully recovered his paper losses. Machiavelli has had a profitable adventure and there has been very good business for lawyers. Who has sustained loss of value resulting from this misadventure? Orange, the children of the neighbourhood, the teachers, the taxpayers, Atlantis and Leonardo. What has Leonardo done or not done which, if done otherwise, would have averted his dilemma? Lastly, the insurance carriers, who appointed legal counsel for Leonardo in defence of a claim against his policy, do not share his interests but are solely interested in minimizing their own financial exposure. They will not pay the costs of initiating legal action to recover Leonardo's losses from his policy.

When we look at worth we need to look at it primarily from the viewpoint of the consumer, in this case the taxpayer. Has the return on the investment been what he was entitled to expect, indeed had secured under contract until it was pulled from under his nose?

Time and posterity will see the events that have been described herein in a dispassionate light, understanding that they will never know the full benefits to the client, children, teachers, neighbourhood, the state and the taxpayer had Leonardo been retained in charge of the project.

To conclude the story; over a long period of time, prior to, concurrent with and subsequent to the events described in this narrative, uncounted numbers of similar dramas unfold until, in time, under the burden of official and unofficial corruption, the democracy that is Atlantis sinks slowly but surely – under the sea.

164 Liberality: Prejudice against Prejudice

Liberal thought is a sophisticated way of embracing experience. It is sophisticated because it attempts to emulate in thought the dynamic processes that prevail in nature. Liberality, as the word suggests, draws its meaning and appeal from the concept of liberty or freedom.

While nothing that we have been able to establish is free in the absolute sense, relative freedom enables flexibility in relation to other things that are in a state of flux. This relative freedom enables the establishment of novel relations that are unattainable or untenable where fixed attitudes prevail. How but in a state of relative freedom from inflexible views can a person prepare for unpredictable change?

Origins

Liberal ideas owe their pedigree to questioning conventional wisdom. From Socrates (who died for this cause) to Jean-Paul Sartre (in his four-volume *Chemins de la Liberté*) men have stood apart from the crowds, often at great risk, to raise questions concerning the just rights of men. It is, before all else, justice that freedom seeks to liberate. The Renaissance, Protestant Reformation and the scientific revolution ushered in the Enlightenment (also known as the Age of Reason) through the course of which a veritable flood of intellectual energy was released, the common theme of which was the pursuit of justice through truth.

Humanity is arguably less than 500 years old if, by humanity, we are referring to a burgeoning rational species distinguished from a few rational specimens. It was from nature that men drew inspiration, and to nature that they turned to justify their liberal ideals. Here at last was a cause to which all could relate in their own interests and in the interests of their society. The socialization of freedom was indeed a cause célébre.

John Locke was to lead the charge of the enlightened brigade for the first century of the Age of Reason through the advancement of the principles espoused in his *Essays on the Law of Nature*, and his earlier-mentioned *Essay Concerning Human Understanding*. That baton passed to Rousseau in the eighteenth century to illuminate and deprecate the extent to which men exhibited emotional insecurity that arose from overbearing social demands to compete, excessive complexity and disparity in wealth, all of which ran counter to more profound interests in understanding the self and its most harmonious place in society as the true bases for human happiness. Today we may, with justification, question whether such states of mind are not even more pronounced than in Rousseau's day. Security in times of aggravating demands demands harmonic accord between peoples more than at any other time. Such harmony can only exist in a state of mutual trust where human relations are founded upon egalitarian principles.

Freedom from Imposition

Every advance in human development can be traced to man's exploitation of the principle of liberty of choice. Art and science as we know them could not otherwise have evolved. Each advance has come as the result of a struggle against entrenched attitudes and behaviour patterns.

The essence of liberal thought is founded upon claims of the prerogative to make individual decisions based upon evidence and circumstance rather than upon predispositions that favour or disfavour any particular group, persuasion or course of action. This presumption is consistent with the declared intentions of the courts in their search for justice. In this sense the liberally inclined live the good life in the present whereas the conservative attempts to live in an extension of the past, while the progressive attempts to live through advancing the future, each according to their predilections of what 'should' constitute the good life.

The Educational Imperative

The emergence of learning as a conscious and expanding source of enlightenment raised the question as to what conditions are necessary for the further development of education. Beyond the obvious physical capacity to collect, store and organize knowledge, the single most significant attribute of learning is the willingness to entertain novel alternatives to those views already held. Underlying willingness is the more profound concept of interest through which potential advantages may be realized. Liberal thought has a duel function in education. One necessarily has to open the mind to new prospects on the one hand, and the new prospects by their nature further open the mind. A natural pattern of growth in understanding emerges through learning. To these ends have been instituted programmes of general education intended to liberalize the way man thinks and communicates. These programmes, known collectively as the Liberal Arts, which include languages, literature, history, philosophy, mathematics and the sciences, lay the foundations for social homogeneity preparatory to teaching and learning professions and trades.

Abstractly, and because of the dynamics of change, it is inconceivable that the level of knowledge of man has reached or will at any time reach a zenith whereby he cannot improve upon his condition through further learning. Only when we consider the condition of individuals can we entertain the idea that we may peak and decline in our abilities to benefit from learning as a natural function of a personal life cycle. Progress is established as the cumulative effect of those increments of knowledge that are discovered by one generation and passed on to the next. Hence the importance of teaching, writing and reading in averting the necessity for each generation to rediscover novelty.

From the foregoing we recognize that Western civilization has benefited immensely from the promulgation of liberal attitudes and institutions that have fostered a great diversity of choice.

The Decisive Animal

Will, resolution and decision form the nub of volition. As we have earlier recognized, the elements of a decision are often not as elementary as one would wish. We are aware of some possibilities and denied knowledge of others. In making judgements we should know whether we are seeking truth or consequences, yet we are unaware of many of the possible outcomes. As a result we enter decision-making at our own risk. It is conscious risk-taking that establishes man as uniquely separate from other animals. As to whether his judgements concur with the absolute truth is absolutely immaterial, but as to whether they translate into positive consequences or not may be life-critical. He has acquired what he himself often attributes to a higher authority: the power to turn events to serve his own interests. The fact that he does not know how his decisions will play out until they are taken weighs heavily upon him. He is damned with responsibilities that are clouded by uncertainty. The burden of responsibilities for poor decisions plagues him with guilt, punishment and remorse.

It is the sunny prospect that spurs him on to make decisions in the understanding that he is free to make his own judgements and that since his beliefs and faith play to the benefit of his doubts he is more likely than not to be an optimist and to decide and act. If one is not making judgements for one's own benefit, nor seeking abstract truths, then presumably one is seeking to decide and act in favour of others. Insofar as it is generally unclear where the distinction between service to self and to others should be drawn, one is naturally attracted to making decisions that favour both causes. That is the liberal inclination, predicated upon optimally fulfilling the individual and collective human potential.

Liberalism in Politics

The cynic views capitalism as the abuse of man by man, communism as the reverse and the proper function of the liberal to be a mediator between the two. The historical view of human nature is of a continuing struggle between self-serving individuals. This view, like its antithesis that societies are by definition cooperative, is too simplistic, failing to recognize that reality comprises an accommodation of both. In 'real life' most of us are a complex mix of conservative, liberal and progressive values. How we sort out when we act in support of each is a function of our experience, circumstance and our intellect.

The liberal poses the proposition that people are best served by a system of government where conflict between the oft-competitive interests is reduced to a minimum. To that end it is government's responsibility to serve each faction honourably. That commitment is constrained by economics. Whereas the individual, through his industry, is the generator of wealth, it is the community that provides the opportunities to do so and thereby to benefit. Each is indebted to the other.

The just apportionment of finite resources thereby becomes central to any government's interest in honouring its obligations. Scarcity is the subject of most of the wrangling most of the time in modern peacetime politics. Strife is generally curable, however, and, contrary to fashionable rhetoric, governments are the agencies most capable of appropriate resolutions.

Try to imagine society without capital or capital without society and one quickly realizes that they are interdependent aspects of human coexistence. It has been the institutionalization of these two weighty topics in the twentieth century, and the consecration of them as 'isms' by their opposite numbers, that has created the schism between them. The cost has been a loss of objectivity essential to balanced relations between capital and society.

The creeds of individualism and socialism are held out by their respective proponents to be political ideals, yet antagonists expose each for their denial of basic freedoms. Thus, freedom has established itself as that common virtue that defines modern politics in all but totalitarian regimes. Inasmuch as differences arise concerning degrees of freedom to be afforded individual and collective interests, any reconciliation between such biased positions necessarily requires to be brokered by those that not only stand politically neutral between these two positions, but for whom human liberty is their principle devotion and basis for compromise.

The most significant shortcoming of political advocacy is failure to recognize the merits in opposing points of view. The function and policy of liberalism in a democracy is to ameliorate the tendencies towards excess that the forces of individualism and collectivism indulge.

A so-called individualist is generally liberal when it comes to claiming his own rights and only conservative where the rights of others are concerned. A collectivist is similarly a liberal advocate of his own rights, becoming more conservative in his commitment to socialism when he addresses society generally. From the above we can recognize that liberal thought is not partisan in the way that other political devotions are, but rather the means by which any open system of social organization can optimize the greatest degree of variety in opportunities to the advantage of individuals and society. It is the liberal quotient in each of us that enables compromise in civilized life, for while we are all individuals, each is a social animal.

The liberal indulgence is eclectic in the sense that one can opt for the best possible world as defined by one's sense of justice without making concessions to political party parsimony. This is because liberal judgements are held tentatively, empirically and undogmatically in the understanding that opinions may change in the light of new evidence, as they do in science but not in theology, where evidence is waived in favour of fervour. Liberal thought and science share a common premise: desire for a natural order wherein authority is self-evident and self-correcting.

Liberal views seldom translate to a liberal majority. A liberal is seldom a party member yet his votes are crucial in the determination of majorities. Liberals generally hold the balance of power in democratic elections. The fact that they often appear not to exploit their pivotal position is a credit to their tolerant attitudes towards the left and right in equal measure. For the liberal the costs and benefits of individualism versus collectivism balance out, and should. In our earlier discussion of politics we defined liberality in the political sense as a high-wire act because the proponent does not rely upon entrenched beliefs for his security, preferring to rely upon his judgement to secure a balance of power. The maintenance of that balance is his political purpose.

While a liberal may be viewed by others as one who vacillates between positions in relation to others on the basis of pragmatism or expedience, the liberal will weight his decisions in relation to broadening rather than narrowing options. He prescribes more rather than less freedom of opportunity, moving to correct injustice caused by disproportionate opposing forces.

Fanaticism is for fanatics who by definition advance the extremities of possibilities. Communism promotes the inequitable distribution of opportunity through the auspices of the state. Capitalism promotes the inequitable distribution of opportunity through the auspices of individual and special interests. Both are in error if, in the conduct of their affairs, they attempt to suppress citizens from attaining 'the separate and equal station to which the laws of Nature and Nature's God entitle them', to quote the US Declaration of Independence; or to deny life, liberty and access to essential health, education and welfare services.

The Society of Free Men

'We have granted to all freemen… for us and our heirs for ever, all the underwritten liberties.' The echoes of Magna Carta still resonate in our minds. What we have been doing for the past 800 years has been to wrangle over who 'free men' should include, and the nature of their entitlements.

In an 'ideal world' presumably all would serve the needs of all others in a kind of liberal Shangri-La, but modern man is intemperate, expending his energies in the pursuit of personal gain at the comparative expense of others. A truly liberated society cultivates the optimization of diversity and variety as an organizing principle. The great majority of people are not born into wide-open spaces and privilege; they are born into congestion and the necessity to reconcile their interests with those of others. Many of our problems arise from our acts of classification (disintegration), much as classification helps to cast light upon their true natures.

We cannot rely upon short-term ideological devotions like market forces or growth to always coincide with collective long-term interests. We must therefore intervene, where there is serious disagreement or degradation, to find solutions that will protect the evolution of life's great variety of formal expressions – in perpetuity. Mankind has always had open prospects for the improvement of his condition.

Prejudice is an attribute of individuals who have reached the limits of their willingness or abilities to entertain prospects for general improvement. This is a function of ageing (dulling of the mind) that will inevitably be surmounted by succeeding generations. Society boils down to a conflict between cooperation and failure to cooperate.

It is the dual realities of the individual and the community that demands just reconciliation. Social systems that show bias in favour of one at the expense of the other are unjust to both.

We cannot help but compare the benefits attributable to the actions of individuals with those due to collective enterprise. All we need to do is to give the correct system the correct credit, or, to the greater credit of our abilities, organize a society that combines the two interests in the manner most beneficial on balance to both; a society that capitalizes on the economical use of things, particularly limited resources, while collaborating to gain advantages for all people. Too often our impulses lead us to polarize our thoughts in a single direction, capitalism or socialism, without carefully considering where the consequences of such impulses may lead us. It has something to do with a lack of will or ability to think fundamentally about what is going on.

Until we have learned as a culture to treat others as we would have them treat ourselves, the two competitive biases will likely continue to grind in cruel opposition.

Notwithstanding claims to the contrary by extremist elements, the Constitution of the United States of America is liberal. It espouses extravagant claims in the Declaration of Independence 'that all men are created equal'; it mandates rule by the majority, it prescribes the twelve-citizen jury system for administering justice and provides free education for its young irrespective of economic means.

If one is fearful of the prospect of a future under liberalism, ask whether there was a period in our evolution in which you would have preferred to live. Would you have been more secure; would you have been better educated; would you have been freer; would you have had greater opportunities; would you have been healthier; always assuming that you were an average man under average circumstances?

The liberal 'freeway' may not lead to everyone's Utopia, but it is surely the path that affords the greatest prospects of advancement for the greatest number. Any system that cannot deliver opportunities for human happiness for the great majority, cannot deliver security for the privileged few.

The history of civilization is the story of the liberation of mankind from the fallacies of conventional dogma and doctrine. As such, liberalism is the most influential creed to infect Western civilization since Christianity, and arguably the only socio-political disposition that offers long-term prospects of human survival.

> 'Give me the liberty to know, to think, to believe, and to utter freely according to conscience, above all other liberties'
>
> John Milton

165 Economy: Resource Distribution and Depletion

We have addressed 'costs' as competitive forces associated with the accrual or loss of benefits; narrow homocentric perspectives of a universal quantitative system of relations. The broader picture, economics, is of a dynamic milieu whereby anything or things may become more or less closely associated with any other thing or things as determined by common standards of measurement.

Nature's Economic Model

Nature is an economist. We might assert that she doesn't seem to care, but if we examine her behaviour carefully we discover, as we did earlier, that she recycles everything. We can construe that the effects of her scrupulous frugality advance her other designs: the distribution of influential parts through the expansion of the universe, the propagation of material socialism through the puff, slosh and grind routines noted earlier, the reliance upon entropy to further the development of new levels of complexity, and her extension through life in general and 'trust' in mankind in particular to accelerate these processes.

The Drifters

Large herds of mammals, flocks of birds and schools of aquatic life migrate. The effect of migration for most of them, most of the time, is to encounter more favourable conditions for feeding, reproduction or hibernation. Inasmuch as they tend to do so relative to changing seasons of the year, their behaviour appears to be habituated. Temperature changes automatically trigger movement towards or away from the sun, the obvious source of direct heat. They are in search of a balance between intolerable extremes of temperature that will not only foster their own wellbeing, but that of that which they like to eat! The extent to which such migratory behaviour has become impressed upon them biologically, as genetic programming, varies by species, but for some it is very difficult to account for why they may venture more than 1,000 miles to congregate at the same locations each winter without invoking foreknowledge.

Butterfly Economics

Generations often elapse between migratory flights of butterflies, and no individuals are known to make long distance two-way trips routinely with the exception of the monarch butterfly.

We might like to argue that survival is the economic principle that propels such behaviour; but we cannot attribute conscious intent to the butterfly. Therefore butterflies do not fly south in the autumn of the northern hemisphere to survive; they survive because they fly south in the autumn. If there was no impression of experience registered by one generation upon another, the probability of congregation in the same locations would be reduced to an extraordinarily degree, while the losses due to random disbursement would be astronomically higher.

The principle of economy in migratory life is therefore two-fold: sufficient numbers must migrate to assure the maintenance of critical stocks of each species. The extent to which their numbers need to exceed the requisite maintenance crew is a function of additional numbers required to offset losses attributable to misinformation, poor conditions and bad luck. Butterflies that fly north into the Canadian winter don't get to tell their stories. We can construe from the above that, while there may be a relation between efficiency and economy, if the living does not actively take care, nothing else will. Caring is a compound function of opportunity, ability, will and discipline.

Economics as a Science

Man, through his observations of the principles affecting supply, has attempted to emulate the mechanisms of nature's exchange processes. Economists examine the implications of rearrangement. Economics is the science of discerning, modelling and testing fluctuations in the supply, distribution and depletion of anything in relation to anything else.

Nature demonstrates its ability to generate more from less: not in the material or energetic sense, but by way of diversity and variety. This is a lesson that man has as yet been unable to fully grasp. He can reconstitute matter and energy, convert from one to the other, and create what has never occurred naturally before; but he seems unable to control the flow of negative impacts that flow from his activity. It is as if his eyes, mouth and feet face in the direction necessary to satisfy his appetite but that he is unaware of, or unconcerned about, collateral consequences that ensue.

Supply

With regard to the nature of 'supply', more is more. Man's interest in economics rests upon a single criterion: is more (of a particular manifestation) a benefit or a detriment to his wellbeing? The pursuit of answers to this question has become his quest. Underlying this question is the need to know. Hence the science of economics, and the economic principles inherent in science generally. The manifold diversity of variety on earth and its seeming infinity of changing relations provide food for thought for the economist and scientist for a long, long time to come.

Distribution

'Distribution' is discriminate in that it conforms to the physical laws respecting the resolution of forces, and indiscriminate in that it has no biased predisposition to serve subjective needs. It is this latter condition that man attempts to modify to his greater advantage. Naturally we tend to focus upon relations between scarcity, abundance and ourselves for those things upon which we depend or value. These are our introductory lessons in economic principles at work. In consequence we have attached a new dimension to nature's economic model, that of 'demand'.

Depletion

'Depletion' is deplorable. It is deplorable in the sense that it works to diminish nature's manifold diversity of variety. But man is single-minded in his interest in pursuing his interests. What we perceive to be in our best interests today may be found tomorrow not to have been so. In short, we too often secure our short-term interests at the expense of our long-term interests. In so doing we work counter to nature's trends towards greater diversity and variety. So the caveat that we need to heed is that our perceptions of self-interest be consistent with our greater interests in survival. Where man is out of sync with nature he is, in the long-term view, out of sync with himself. That is psychotic. That is madness! If he is not committed, he should be!

Demand

'Demand' (meaning insistence by command) is an invention of man. It is subject to his limited grasp of possible causes and probable consequences. In generating demands and taking actions to satisfy them, he diverts the course of events that would otherwise occur under the auspices of nature-absent-man. To the extent that he causes no irrevocable harm to nature's manifold variety of viable forms and systems, nature does not intervene. To the extent that man irrevocably detracts from nature's diversity, nature penalizes man by removing from the realm of possibilities any further acquaintance with, or potential benefits to be drawn from, such phenomena that are thereby subtracted from nature's inventories.

Retribution for Excess

Dodo birds were no dodos! It was the early settlers on the island of Mauritius in the seventeenth century that knew not what they were doing when they decimated the species into extinction. If one decimates without remission, one reduces say 100 dodos to 90, 90 to 81 and 81 to 72.9. For obvious reasons we must discount decimals of decimated dodos. From 72 we proceed to 64, and on to 57, 51, 45, 40, 36, 32, 28, 25, 22, 19, 17, 15 (all the time getting a 10 per cent smaller yield from the hunt), and beyond to 13, 11, 9, 8, 7, 6, 5, 4, 3, 2; and finally a single irreducible, irreproducible bird. From this point onward 'Thanksgiving' dinners were never quite the same! The thankless settlers abandoned the island.

The fate of the Moa birds of New Zealand was much the same, though the identity of the real 'dodos' in this case is less clear.

The Lesson of the Lesson

The moral of the message is not to shoot oneself in the left foot in the understanding that one has another (the right), since without a pair, one may despair of making any significant advances. The word 'moral' is chosen purposefully to underline the proposition that shooting dodo birds, feet, or in any other way unnecessarily impairing the 'natural' course of development of diversity and variety is 'immoral'. For the god-fearing, God and Nature may be considered to speak with equal authority.

Margins of Tolerance

The corollary to this discussion is the necessity for strict boundaries to be drawn on a case-by-case and place-by-place basis to limit the extent to which man may be permitted to undermine the integrity of natural living systems. Nor should vital resources upon which life directly or vicariously depends be expended at greater rates than can be replenished. Any such commitment should also provide for restraint, restoration and restitution.

Money isn't Anything

The principle behind economic variation in human affairs is purportedly the relation between demand and supply, but when it comes to money this nuance breaks down. The accrual of money is the exception. An increase in the supply of money seems to accelerate demand. There is never a glut, never too much. Where money is concerned, an increase in supply stimulates increases in expectation. If such expectations are not satisfied, the negative impacts are more detrimental to those effected than if they had no such false expectations to start with. In the context of economics – being concerned with supply, distribution and depletion of anything in relation to anything else – money isn't 'anything', it is a facsimile of almost everything, thus explaining its maverick behaviour as an unreliable economic indicator – unless one is tracking the supply, distribution and depletion of facsimiles!

Applied Economics

In the world of what we shall call here 'applied economics', concepts of value have been reduced to relative prices, the word 'price' replacing 'value' in our vernacular. Insofar as exchange rates for goods and services are reduced to a common frame of reference, money, it has been irresistible to entertain the notion that all values may be so reduced.

Economics is a social science. As such, distinctions need to be made between essential and non-essential production, distribution and consumption to the inurement of social wellbeing over time. We do this willingly at times of impending disaster. We need to focus on long-range potentials for disaster so that we can mitigate threats by minimizing adverse causes and effects. This requires planning, a discipline that often finds itself at cross purposes with the short-term expedient agendas of politics.

Economics places limits on the possible. Severe economic limitation has the effect of denying alternative outcomes; indeed denial due to limitations of existing systems is the most powerful promoter of social, technological and institutional change. Wilful denial of otherwise legitimate economic opportunities is a form of enslavement.

Autonomic Economics

Economic forces are the purveyors of life. They nourish all aspects of progenitive vitality. Supply and distribution are effected through the respiratory, digestive, pulmonary and neural systems in animals, or their equivalence in plants. All of these systems are programmed to function automatically. Insofar as commercial economics deals with the transference of numeric values, the neural system presents the best analogue for how we might aspire to telegraph and control the supply, distribution and depletion of economic values within society. To these ends we have harnessed computer technology to do much of our work for us, the ultimate goal being to rise above the state of competition for selfish gain to a state where the security of the State, and through it the individual, could be assured through the supply and distribution of its requirements (demands), substantially in an automatic mode.

What the rich have now is semi-automatic wealth generation. The process will in time flow to the relief of public (social) needs. What does this do to the nineteenth-century work ethic? It places a great responsibility upon the individual to use his assets wisely, not the least of which is the time that he gains through relief from the need to generate wealth. The secret side of economics is the function it performs in reflecting self-worth. What you do with what you've got is who you are.

If we take the high view, that of God, nature's wondrous processes may be seen to be autonomic (i.e. independent of volition). They utilize, conserve and generate constant amounts of energy, they maintain a steady balance of matter *in toto*, and they evidently require no wilful direction from man or God.

When we have learned how to create wealth and to govern fluctuations in the economy in large measure autonomically, what logically comes next? At such a time we shall be truly liberated to focus our attentions upon higher purposes: the cultivation of human affections.

Index